574 POT

STUDENT UNIT GUIDE

NEW EDITION

AQA A2 Biology Unit 4
Populations and Environment

Steve Potter and Martin Rowland

 PHILIP ALLAN

Philip Allan Updates, an imprint of Hodder Education, an Hachette UK company, Market Place, Deddington, Oxfordshire OX15 0SE

Orders
Bookpoint Ltd, 130 Milton Park, Abingdon, Oxfordshire OX14 4SB
tel: 01235 827827
fax: 01235 400401
e-mail: education@bookpoint.co.uk
Lines are open 9.00 a.m.–5.00 p.m., Monday to Saturday, with a 24-hour message answering service. You can also order through the Philip Allan Updates website: www.philipallan.co.uk

© Steve Potter and Martin Rowland 2012

ISBN 978-1-4441-5292-0

First printed 2012
Impression number 5 4 3 2
Year 2017 2016 2015 2014 2013

Cover image: fusebulb/Fotolia

Printed in Dubai

Hachette UK's policy is to use papers that are natural, renewable and recyclable products and made from wood grown in sustainable forests. The logging and manufacturing processes are expected to conform to the environmental regulations of the country of origin.

Contents

Content Guidance

Questions & Answers

Getting the most from this book

Examiner tips

Advice from the examiner on key points in the text to help you learn and recall unit content, avoid pitfalls, and polish your exam technique in order to boost your grade.

Knowledge check

Rapid-fire questions throughout the Content Guidance section to check your understanding.

Knowledge check answers

1 Turn to the back of the book for the Knowledge check answers.

Summary

Summaries

● Each core topic is rounded off by a bullet-list summary for quick-check reference of what you need to know.

Questions & Answers

Exam-style questions

Examiner comments on the questions
Tips on what you need to do to gain full marks, indicated by the icon ⓔ.

Sample student answers
Practise the questions, then look at the student answers that follow each set of questions.

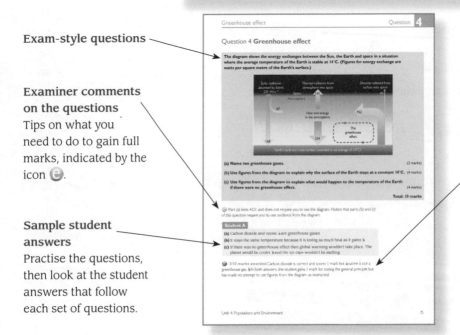

Examiner commentary on sample student answers
Find out how many marks each answer would be awarded in the exam and then read the examiner comments (preceded by the icon ⓔ) following each student answer. Annotations that link back to points made in the student answers show exactly how and where marks are gained or lost.

AQA A2 Biology

About this book

This guide will help you to prepare for **BIOL4**, the examination for **Unit 4: Populations and Environment**, of the AQA A-level Biology specification. Your understanding of some of the principles listed at the end of Units 1 and 2 may be re-examined here as well.

The **Content Guidance** section covers all the facts you need to know and concepts you need to understand for BIOL4. In each topic, the concepts are presented first. It is a good idea to make sure you understand these key ideas before you try to learn all the associated facts. The Content Guidance also includes examiner tips and knowledge checks to help you prepare for BIOL4.

The **Question and Answer** section shows you the sorts of question you can expect in the unit test. It would be impossible to give examples of every kind of question in one book, but these should give you a flavour of what to expect. Each question has been attempted by two students. Their answers, along with the examiner's comments, should help you to see what you need to do to score a good mark — and how you can easily *not* score a mark even though you might understand the biology.

What can I assume about this book?

You can assume that:

- the basic facts you need to know and understand are stated explicitly
- the major concepts you need to understand are explained clearly
- the questions at the end of the guide are similar in style to some of those that will appear in the BIOL4 unit test
- some of the questions test aspects of *How Science Works*
- the answers supplied are the answers of A2 students
- the standard of the marking is broadly equivalent to the standard that will be applied to your answers

So how should I use this book?

The guide lends itself to a number of uses throughout your course — it is not *just* a revision aid. You can use it:

- to check that your notes cover the material required by the specification
- to identify your strengths and weaknesses
- as a reference for homework and internal tests
- during your revision to prepare 'bite-sized' chunks of related material, rather than being faced with a file full of notes

You could use the Question and Answer section to:

- identify the terms used by examiners in questions and what they expect of you
- familiarise yourself with the style of questions you can expect
- identify the ways in which marks are lost as well as how they are gained

Develop *your* examination strategy

As an A2 student, you will by now have realised that you need to do more in order to prepare for a unit test than simply reading your notes or a textbook. You need to develop and maintain the skills that the examiners will test in BIOL4. You also need to know what sort of questions the examiners will ask and where to find them in BIOL4. If you prepare in this way, you will not be surprised by BIOL4 and will have a strategy for answering the questions. But, be warned, developing your strategy is a highly personal and long-term process; you must start it at the beginning of your A2 course.

Things you *must* do

- Clearly you must know and understand the topics of ecology, photosynthesis, respiration, genetics and natural selection. If you don't, you cannot expect to get a good grade. This guide provides a succinct summary of these topics.
- Prepare for an *approximate* mark breakdown for these topics in BIOL4 to be:
  - ecology 40 marks
  - photosynthesis and respiration 23 marks
  - genetics 12 marks
- Understand that the weighting of assessment objectives that examiners *must* use in BIOL4 has a stronger emphasis on AO2 and AO3 than your AS units. Examiners have designed BIOL4 with the approximate balance of marks shown in the table.

Assessment objective	Brief summary	Marks in BIOL4
AO1	Knowledge and understanding	22
AO2	Application of knowledge and understanding	35
AO3	How Science Works	18

- Understand where in BIOL4 different types of question occur. For example, the final question will be a structured essay, usually in three parts worth 5 marks each, testing mainly AO1 (this means that 15 of the 22 marks testing AO1 are in this last question). The penultimate question, worth 15 marks, will contain most of the AO3 marks and could include biological content from anywhere within the AS specification or Unit 4.
- Use questions from past papers, from your textbook or from websites to maintain all the skills that examiners must test and which you should have begun to develop in your AS course. Remember, practice now means you will be more confident, and more in control, when you sit BIOL4.

Content Guidance

The Content Guidance section is a guide to the content of **Unit 4: Populations and Environment**. It contains the following features.

Key concepts you must understand

Whereas you can learn facts, these are ideas or concepts that may form the basis of models that we use to explain aspects of biology. You can know the words that describe a concept like ecological niche or oxidative phosphorylation, but you will not be able to use this information unless you really understand what is going on. Once you genuinely understand a concept, you will probably not have to learn it again.

Key facts you must know and understand

These are exactly what you might think: a summary of all the basic knowledge that you must be able to recall and show that you understand. The knowledge has been broken down into a number of small facts that you must learn. This means that the list of 'Key facts' for some topics is quite long. This approach, however, makes quite clear *everything* you need to know about the topic.

Summary

This describes the skills you should be able to demonstrate after studying each related topic. These include the skills associated with the assessment objectives that examiners will ask you to demonstrate in the BIOL4 unit test.

Content Guidance

Populations and the factors that affect them

Populations and ecosystems

Key concepts you must understand

Population

A population is all the organisms of **one** species found in a particular habitat at a particular time.

Community

A community is all the populations found in a particular habitat at a particular time.

Ecosystem

An ecosystem is a self-contained unit consisting of a community and its non-living environment. Ecosystems vary greatly in size. A garden pond could be considered an ecosystem, as could a tropical rainforest.

Habitat

A habitat is the place in an ecosystem where a community of organisms is found. A rocky shore and the human skin are examples of habitats in which different communities live.

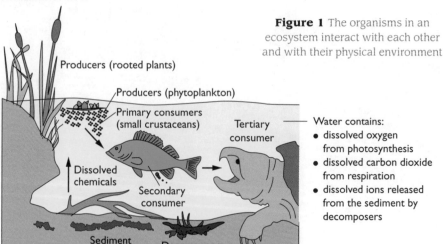

Figure 1 The organisms in an ecosystem interact with each other and with their physical environment

Producers (rooted plants)

Producers (phytoplankton)

Primary consumers (small crustaceans)

Tertiary consumer

Dissolved chemicals

Secondary consumer

Water contains:
- dissolved oxygen from photosynthesis
- dissolved carbon dioxide from respiration
- dissolved ions released from the sediment by decomposers

Sediment

Decomposers

Knowledge check 1

Your skin can be regarded as a habitat. Explain why.

Niche

A niche is not just a place where a species is found; that is the habitat. A niche is a description of how the species functions in that habitat. A description of a niche takes into account:

- **abiotic conditions**, i.e. non-living aspects of the environment like the temperature a species can tolerate
- **biotic conditions**, i.e. biological aspects such as the availability of a suitable food source or the presence of competitors

When the two environmental resources (one biotic and one abiotic) are combined, there is only a small area of overlap between the two species. This is represented by region 5 in the diagram.

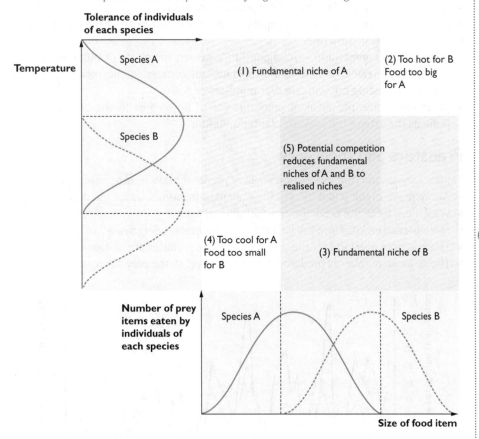

Figure 2 Different organisms have different niches

Factors affecting population size

Key concepts you must understand

Population size is dependent on the availability of resources. If resources are unlimited, a population is capable of growing exponentially. Usually, resources are not unlimited; a combination of abiotic and biotic factors restricts population growth.

Key facts you must know and understand

Abiotic factors are the non-living components of an ecosystem, and include:

- carbon dioxide concentration in the atmosphere — affects populations of photosynthesisers
- mineral ion availability in the soil — affects populations of photosynthesisers
- light intensity — affects populations of photosynthesisers and some animal populations
- water availability — affects all populations
- temperature — affects all populations

Biotic factors are the biological components of an ecosystem, and include:

- disease-causing organisms — slow population growth
- intraspecific competition — competition between members of the same species slows population growth
- interspecific competition — competition between members of different species. This usually results in one of the species out-competing the other or the two species coexisting but with smaller numbers of both
- predation — the presence of predators (or of herbivores in the case of plants) reduces the size of the prey (or plant) population

Knowledge check 2

In woodland, bluebells and trees compete for light. Identify precisely the biotic and abiotic factors in the previous sentence.

Predators and their prey

Figure 3 shows that the population sizes of predator and prey are interdependent.

- An increase in the population of the prey means more food for the predators.
- With more food, the predator population increases.
- More predators kill more prey, so the prey population decreases.
- There is now less food for the predators, so the predator population decreases.
- The reduced number of predators kills fewer prey, so the prey population increases.

Knowledge check 3

Look at Figure 3. Explain why (a) the rise and fall in the predator population always lags behind that in the prey population and (b) the prey population is larger than the predator population.

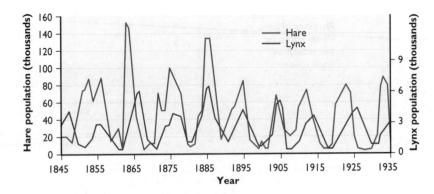

Figure 3 Changes in the population size of snowshoe hare and lynx from 1845 to 1935

Interspecific competition

Paramecium is a type of ciliated protoctist that lives in water. Figure 4 shows that when three species of *Paramecium* are grown separately, each reaches a maximum population size.

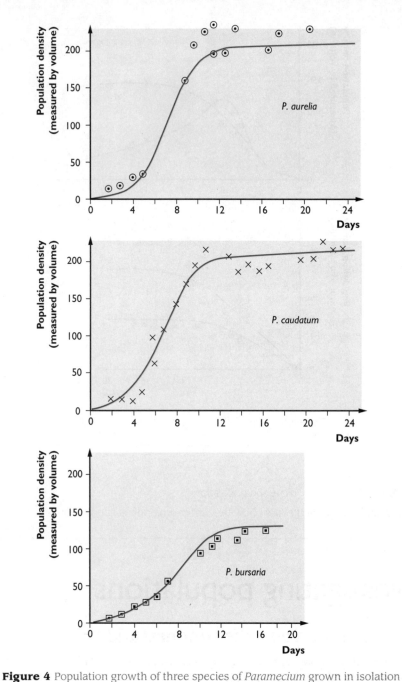

Figure 4 Population growth of three species of *Paramecium* grown in isolation

Figure 5 shows what happens when two species of *Paramecium* are grown together. When *P. caudatum* and *P. aurelia* are grown together, *P. aurelia* out-competes *P. caudatum*, which becomes locally extinct.

When *P. caudatum* and *P. bursaria* are grown together, the species coexist, but with smaller populations than when grown alone.

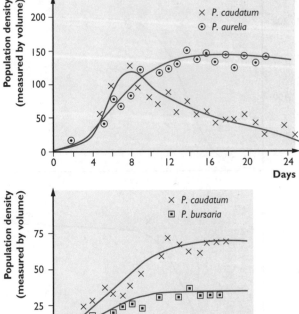

Knowledge check 4

(a) Suggest **one** factor that limits population size of *P. caudatum* in Figure 4.

(b) Name **one** factor that limits population size of *P. caudatum* in Figure 5.

Figure 5 Interspecific competition in *Paramecium*

Summary

After studying this topic, you should be able to:

- define the terms community, ecosystem, habitat, niche and population
- distinguish between abiotic and biotic factors and explain how they act to limit population growth

Investigating populations

Key concepts you must understand

It is usually impossible to count all the organisms in a population, so we almost always *estimate* the size of a population.

To make an estimate, we count the numbers of organisms in a *sample* of the population. A sample is a subset of the population. For the estimate to be reliable, the sample should be representative of the population as a whole.

We cannot *choose* a sample, because that would introduce **bias**. The bias would be towards our opinion of what a typical sample would be. The sample should, therefore, be a **random sample**.

The numbers of any one species in an ecosystem can be estimated using:
- **quadrats** — for relatively stationary organisms
- **mark–release–recapture techniques** — for mobile species

Key facts you must know and understand

Using random quadrats to estimate population size

You would use this method if you suspected there was no environmental change in the area, for example in a playing field.
- Divide the area into a grid and give the grid coordinates, as in Figure 6.
- Use the random number generator on a calculator to produce a pair of coordinates, for example A5, B6 (this pair would define the top left-hand corner of the shaded square in Figure 6).
- Repeat this procedure until you have identified a sufficient number of samples to make your data reliable.
- Place a quadrat frame with its top left-hand corner on the intersection of one pair of coordinates.
- Estimate the abundance of the different organisms in each quadrat using one of the following methods.
  - Count the numbers of each organism in each quadrat.
  - Calculate the **percentage frequency** of occurrence. This is done by recording presence or absence in each quadrat and converting the number of occurrences to a percentage (e.g. a species that occurs in eight out of ten quadrats has a frequency of 80%).
  - Calculate the **percentage cover**. This is done either by making a crude estimate of the percentage of each quadrat covered by the species or by using a quadrat that is subdivided into smaller squares — a 'gridded' quadrat, as shown in Figure 7.
    If the organism covers 7 squares in Figure 7, the percentage cover is 7 × 4% = 28%.
    If the organism covers 8 squares and 2 part-squares in Figure 7, a total of 9 squares is a reasonable estimate. This would give 9 × 4% = 36% cover.
    These quadrats can also be used to estimate frequency. Suppose an organism occurs in 12 of the 25 squares (the actual number in each small square does not matter). The frequency of occurrence is:

$$\frac{12 \times 100}{25} = 48\%$$

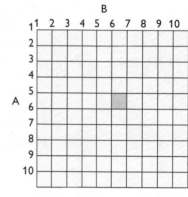

Figure 6 An area divided into a grid with coordinates

Knowledge check 5

How could you decide when you had a sufficient number of samples to make your data reliable?

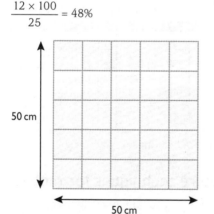

This quadrat has 25 small squares. Each square represents $^1/_{25}$ or 4% of the total area.

Examiner tip

Do not describe throwing quadrat frames over your shoulder as a method for estimating population size. It will not gain marks but will waste valuable exam time.

Figure 7 A 'gridded' quadrat

- Find the mean abundance per quadrat.
- Knowing the area of the quadrat, convert the mean abundance per quadrat to mean abundance per standard area, for example per m^2.

Using quadrats along a belt transect to estimate population size

You can see straight away that some habitats are not uniform. A rocky shore is a good example; you can see bands with different seaweeds running across it. Figure 8 represents a field in which there are clearly different areas. The upper part of Figure 8 shows that if taking random samples, as described above, you might miss sampling one of these areas. We overcome this by taking samples systematically along a **transect**, as shown in the lower part of Figure 8.

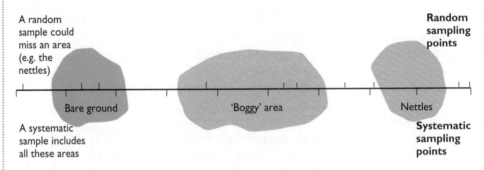

Figure 8 Systematic and random sampling along a transect

Figure 9 shows how we would start to use quadrats along a transect.
- Lay a tape measure across the sample area.
- At regular intervals (every 4 m in Figure 9) lay five quadrats to one side of the tape (always the same side as shown in Figure 9).
- Estimate the abundance of different organisms at each sampling point along the transect using one of the methods described for random sampling above.

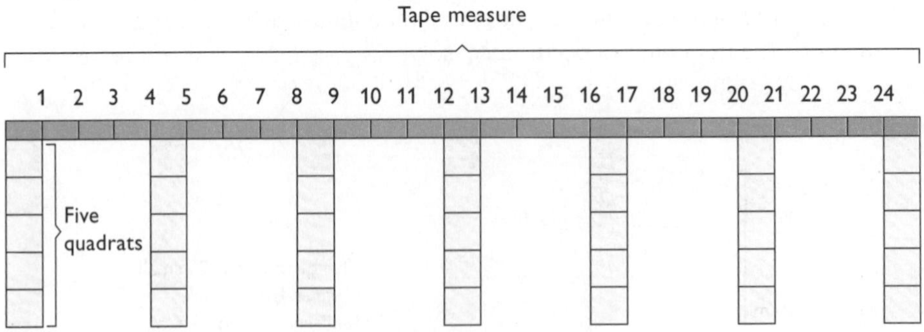

Knowledge check 6

Would you use random sampling or a belt transect to investigate the plant populations on a motorway verge? Explain your answer.

Tape measure

1 2 3 4 5 6 7 8 9 10 11 12 13 14 15 16 17 18 19 20 21 22 23 24

Five quadrats

Figure 9 A belt transect

Using mark–release–recapture techniques for mobile species

The methods described above are fine if you are investigating plants or sessile animals. The mark-release-recapture method is suitable for animals that are highly mobile.

- Collect a sample of the animals from the area and count them (N_1).
- Put a small mark in an unobtrusive place on each animal.
- Release the marked animals and allow time for them to disperse among the population.
- Collect a second sample and count both the total size of the sample (N_2) and the number that are marked (n).
- Estimate the population size using the formula
$$\frac{N_1 \times N_2}{n}$$

For example, if you caught, marked and released 50 woodlice and later caught 40 of which 10 were marked, you would estimate the population as $(50 \times 40)/10 = 200$.

Looked at another way, in the second sample of 40, 10 (a quarter) were marked, so we assume that one-quarter of the entire population is marked. So 50 (the number originally marked) represents a quarter of the population. The population is therefore 200.

Examiner tip

Many candidates show in exams that they find it easier to work out population size by remembering that:

$$\frac{\text{number of individuals released}}{\text{total population}} = \frac{\text{number of marked individuals in 2nd sample}}{\text{total number in 2nd sample}}$$

When using this technique we assume that, during the time the marked individuals disperse among the population:
- there is no migration
- there are no births or deaths
- there is random mixing of the marked and unmarked individuals
- marking does not affect behaviour

Knowledge check 7

A student captured and marked 10 snails and released them back in his garden. The next day he collected 12 snails of which 3 were marked. (a) How many snails lived in his garden? (b) What assumptions have you made in finding your answer?

Human populations

Key concepts you must understand

Excluding migration, any change in the size of a human population is a result of the difference between numbers born and numbers dying. There are three possibilities:
- births > deaths; the population increases
- births < deaths; the population decreases
- births = deaths; the population remains the same size
- A change in population size is usually expressed as a **population growth rate** or **rate of natural increase**. This is determined from:
  - **birth rate**: the number of births per 10000 people
  - **death rate**: the number of deaths per 10000 people

Growth rate (rate of natural increase) = birth rate – death rate

Worked example 1

Suppose the birth rate is 14 per 10000 and the death rate is 8 per 10000.

The population growth rate is $14 - 8 = 6$ per 10000 or +0.06%.

The population is increasing.

Worked example 2

Suppose the birth rate is 14 per 10 000 and the death rate is 16 per 10 000.

The population growth rate is 14 – 16 = –2 per 10 000 or –0.02%.

The population is decreasing.

The changes in birth and death rates shown in Figure 10 can be used to decide whether a population is growing, shrinking or remaining static.

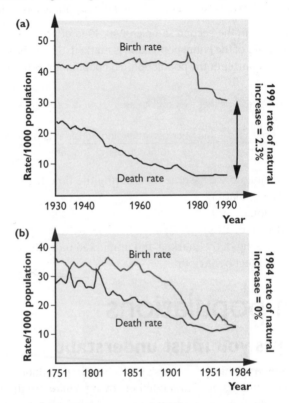

Figure 10 (a) Birth and death rates in Mexico 1930–1990
(b) Birth and death rates in Sweden 1751–1984

In Mexico:
- birth rate was always higher than death rate, so the population was increasing
- both birth rate and death rate were falling, but birth rate was falling more slowly
- the gap between birth and death rates was getting wider
- the rate of natural increase (population growth rate) was increasing

In Sweden:
- for most of the time, birth rate was higher than death rate, so the population increased for most of the time
- both birth rate and death rate were falling, but birth rate was falling faster
- the gap between birth and death rates was getting narrower, and was zero in 1984
- the rate of natural increase (population growth rate) was decreasing and was zero in 1984: after this point the population remained the same size

Examiner tip
Be careful not to confuse population size with population growth rate. In Figure 10(b), although the growth *rate* (rate of natural increase) of the Swedish population was decreasing, the population *size* increased until the growth rate was zero.

Figure 11(a) shows how most populations are restricted by **limiting factors**. By modifying our environment, humans have been able to change these limiting factors. As a result, our world population has achieved the rapid growth shown in Figure 11(b).

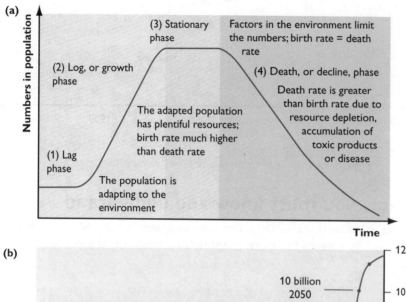

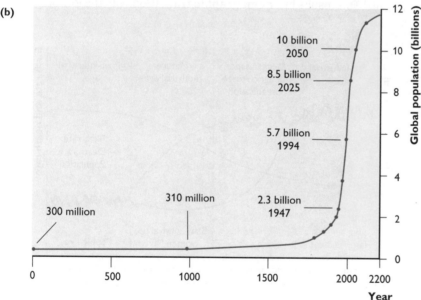

Figure 11 (a) The typical growth phases of a population
(b) The growth of the human population over the past 2000 years together with the projected growth for the next 200 years

The growth of the human population has mainly resulted from reduction in death rates due to:

- increased quality and quantity of food available
- improved sanitation
- improved medical care

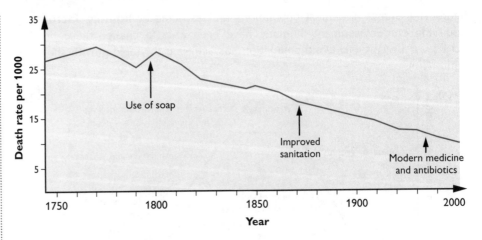

Figure 12 Decline in death rate in Sweden from 1750 to 2000

Key facts you must know and understand

Historically, we know that the size of human populations is affected by technological developments. This results in a **demographic transition** with the four stages shown in Figure 13.

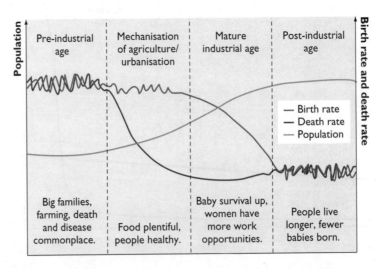

Figure 13 The stages of the demographic transition

In the second and third stages of the demographic transition, death rates fall before birth rates and so the population still increases. In the final stage, birth rates and death rates are low and the population becomes stable.

Most developed countries are in this final stage of the demographic transition; most developing countries are in one of the two middle stages. As a result, most population growth is occurring in developing countries.

During demographic transition, the relative number of young and old people changes. These changes are best shown in **age pyramids**. Figure 14 shows age pyramids for

AQA A2 Biology

rapidly expanding populations, slowly expanding populations, stable populations and declining populations.

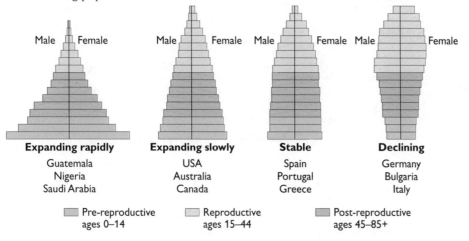

Figure 14 Age pyramids

A broad base to a population shows that large numbers of children are being born and that the population is increasing. This structure is common in developing countries.

Similar numbers in all the age groups until old age is reached shows that the number being born is approximately equal to those dying and the population is stable. This is true of many developed countries, where life expectancy is also greater.

Knowledge check 9

What can you conclude from Figure 14 about (a) infant mortality in stable populations and (b) female mortality?

Summary

After studying this topic, you should be able to:

- describe how to estimate abundance using random quadrats, quadrats along a transect and the mark–release–recapture method and explain when it is appropriate to use each method
- explain the assumptions on which the mark–release–recapture method is based and use given data to calculate the size of a population
- relate changes in the size and age structure of human populations to different stages in demographic transition
- calculate population growth rates from given data

The transfer of energy within organisms

The structure of ATP

Key concepts you must understand

Figure 15 shows that a molecule of ATP contains:
- one molecule of adenine (one of the nitrogenous bases found in DNA and RNA)
- one molecule of ribose (the pentose found in RNA)

(The combination of adenine and ribose is called adenosine.)

- three phosphate groups

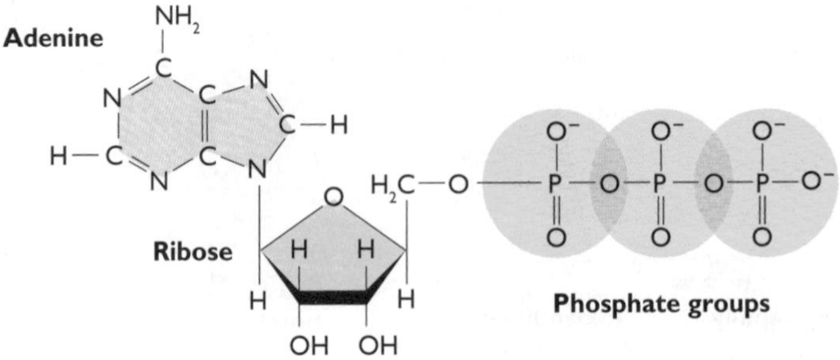

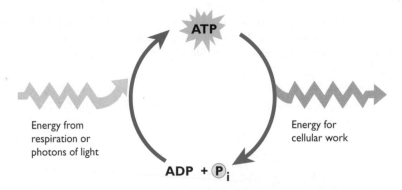

Figure 15 Structure of ATP (adenosine triphosphate)

The third (outermost) phosphate group can be split off from the rest of the molecule. When this happens, energy is released that can be used to do useful work in the cell. Splitting off the third phosphate produces ADP (**a**denosine **dip**hosphate) and P_i (inorganic phosphate). This one-step reaction is catalysed by the enzyme **ATPase**.

ADP and P_i can be joined again to make ATP, which requires an *input* of energy.

ATP

Energy from
respiration or
photons of light

Energy for
cellular work

ADP + P$_i$

Figure 16 The interconversion of ATP, ADP and P_i

The energy required to synthesise ATP from ADP and P_i can come from:

- cellular respiration
- the transduction of light energy in photosynthesis

Key facts you must know and understand

The main ways in which the energy released by the hydrolysis of ATP is used are shown in Figure 17.

ATP is the molecule that releases energy to drive biological processes. It is said to be **coupled** to these processes. It is an ideal molecule for this function because:

- energy is released from the molecule quickly, in a single-step hydrolysis reaction
- energy is released in small amounts that are matched closely to the amounts used in coupled cellular reactions
- the molecule is moved around easily within the cell, but cannot leave the cell

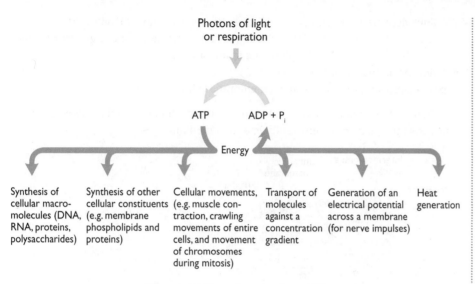

Figure 17 Uses of energy from ATP

Examiner tip

Do not to tell examiners that energy is produced; it is **released** when ATP is hydrolysed.

Photosynthesis: harnessing light energy

Key concepts you must understand

Photosynthesis is a process with two main stages: the **light-dependent reaction** and the **light-independent reaction**.

- In the light-dependent reaction, light energy is absorbed by pigments (that include **chlorophyll**) in the **chloroplasts** and is used to synthesise ATP and **reduced NADP**.
- In the light-independent reaction, the ATP and reduced NADP from the light-dependent reaction are used to drive reactions that result in the synthesis of glucose.

NADP can act as an electron carrier. When it accepts electrons, we say it has been reduced.

In the light-independent reaction, ATP and reduced NADP are used in the following ways:

- ATP supplies energy to drive **endergonic** (energy-requiring) reactions and provides phosphate groups to phosphorylate other compounds.
- Reduced NADP gives up its electrons, which are used to reduce another compound, glycerate 3-phosphate.

Key facts you must know and understand

How chloroplasts are adapted for their function

The structure of chloroplasts allows the two reactions of photosynthesis to occur efficiently. Figure 18(a) shows a single chloroplast.

Examiner tip

The final question in BIOL4 is a structured essay. Here, you might be asked to write an account of the light-dependent reaction and/or light-independent reaction for 5 or 6 marks. The bullet points in section 3.4.3 of the specification will form the mark scheme for such a question.

The light-dependent reaction takes place in the membranes of **thylakoids**.
- The thylakoids are stacked into structures called **grana** (singular **granum**), which maximises the ability of the pigments to absorb light energy.
- Chlorophyll and other light-absorbing pigments are organised into clusters, called **photosystems**, on the thylakoid membranes shown in Figure 18(c).

The light-independent reaction takes place in the liquid stroma. Many chemical reactions take place most efficiently in a fluid medium.

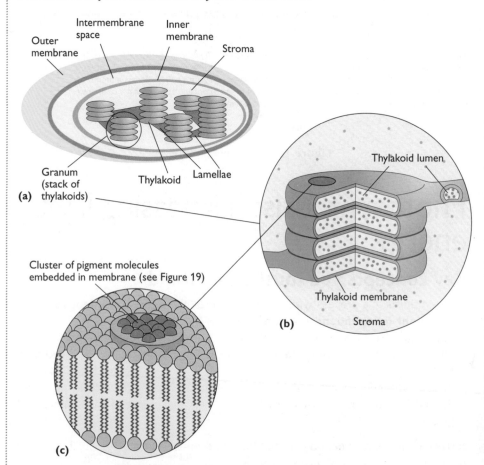

Figure 18 (a) The structure of a chloroplast, (b) A section through a single thylakoid, (c) Pigments in a thylakoid membrane

Figure 19 shows part of a single photosystem. It shows how energy from **photons** of light is absorbed by pigment molecules and transferred to a chlorophyll molecule within a **reaction centre**. This chlorophyll molecule is linked to a **primary electron acceptor**. Electrons (e⁻) from chlorophyll molecules are taken up by this primary electron acceptor.

The primary electron acceptor transfers electrons from the chlorophyll to a chain of molecules called an **electron transfer chain**. As the electrons pass along the electron transfer chain, energy is released, which is used to synthesise ATP from ADP and P_i. At the end of the chain, the electrons reduce NADP.

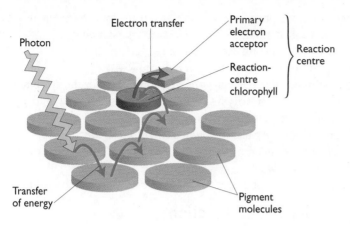

Figure 19 A photosystem

Knowledge check 10

The electron transfer chain involves a series of reduction–oxidation (redox) reactions. Explain this statement.

The light-dependent reaction

The light-dependent reaction takes place in the **grana**. There are two different types of photosystems in the membranes of the grana that absorb different wavelengths of light. They are called **photosystem I** and **photosystem II**.

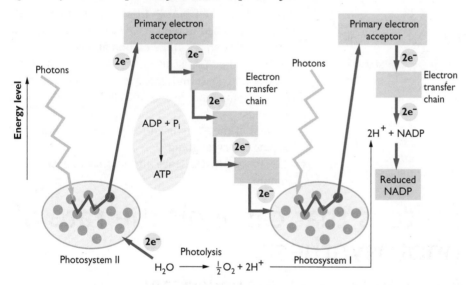

Figure 20 Light-dependent reaction

- Light energy is absorbed by pigments in photosystem II and passed to the chlorophyll in its reaction centre.
- The absorbed energy raises the energy levels of electrons in the chlorophyll molecule so that a pair of electrons leaves the chlorophyll molecule.
- These electrons are taken up by the primary electron acceptor and then passed along an electron transfer chain. As they pass along the chain, energy is released and used to generate ATP from ADP and P_i.
- The electrons are transferred to the chlorophyll in photosystem I.
- When further light energy is absorbed by photosystem I, two more electrons are excited and escape from this chlorophyll molecule.

- The electrons from photosystem I are taken up by a different primary electron acceptor and passed along a second electron transfer chain.
- At the end of this second electron transfer chain, the electrons (e^-) react with hydrogen ions (H^+) and NADP to form reduced NADP.
- The ATP and reduced NADP leave the granum and enter the stroma.
- In a process called **photolysis**, water molecules are split into hydrogen ions, electrons and oxygen atoms.
  - The hydrogen ions are used to reduce NADP, as above.
  - The electrons are taken up by the chlorophyll in photosystem II, replacing those it lost.
  - The oxygen atoms combine to form molecular oxygen, which leaves the chloroplast.

> **Knowledge check 11**
>
> Name (a) the products of the light-dependent reaction and (b) a waste product of the light-dependent reaction.

The light-independent reaction

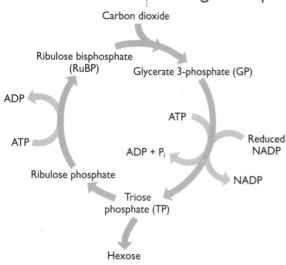

Figure 21 Light-independent reaction (the Calvin cycle)

This takes place in the stroma.

- Carbon dioxide enters the stroma and reacts with **ribulose bisphosphate** (**RuBP**), which has five carbon atoms. This reaction produces two molecules of **glycerate 3-phosphate** (**GP**), which have three carbon atoms. The reaction is catalysed by the enzyme **Rubisco**.
- ATP and reduced NADP from the light-dependent reaction are used to reduce GP to **triose phosphate** (**TP**), which also has three carbon atoms.
- Some of the TP is used to synthesise hexoses (6-carbon sugars) and, from them, starch and other organic compounds (such as cellulose, amino acids and lipids).
- The rest of the TP is used to synthesise more RuBP, keeping the cycle going.
- ADP, P_i and NADP leave the stroma and enter the granum.

> **Knowledge check 12**
>
> What do ATP and reduced NADP contribute to the reaction by which GP is converted to TP in the light-independent reaction?

Factors that can limit the rate of photosynthesis

Key concepts you must understand

The main factors that influence the rate of photosynthesis and the way in which they influence the rate are shown in the table below.

Factor	Influence on the rate of photosynthesis
Light intensity	Light energy 'drives' the light-dependent reaction. Increasing the intensity of light increases the rate of this reaction.
Carbon dioxide	Carbon dioxide reacts with RuBP in the light-independent reaction. Increasing the concentration of carbon dioxide increases the rate of this reaction.
Temperature	Many of the reactions of photosynthesis are controlled by enzymes. Increasing the temperature to the optimum temperature for the enzymes increases the rate of photosynthesis. Above this temperature, enzymes will start to denature and the process will quickly slow down.

When several factors influence the rate of a process, their effects combine and the factor that is present in the 'least quantity' has the most influence on the overall rate. It is the **limiting factor**.

Examples of limiting factors

On a bright, sunny day in January, with a temperature of +1°C, the low temperature probably limits the rate of photosynthesis. The enzymes are working well below their optimum temperature.

On a bright, sunny day in July, with a temperature of +28°C, the concentration of carbon dioxide may limit the rate of photosynthesis. The light intensity is high, and the temperature is close to the optimum for most plant enzymes in temperate countries. With more carbon dioxide, the process would probably proceed faster.

If light intensity is the factor that limits the rate of photosynthesis, then increasing the light intensity will increase the rate of photosynthesis. Increasing the concentration of carbon dioxide (when light is limiting) will have no effect.
- Increasing a non-limiting factor has no effect on the rate of photosynthesis.
- Increasing a limiting factor increases the rate of photosynthesis until some other factor becomes limiting (see Figure 22).

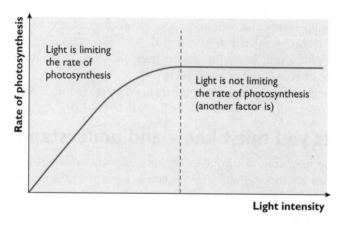

Figure 22 Effect of increasing light intensity on the rate of photosynthesis

Figure 23 shows that if we now introduce the effect of carbon dioxide concentration as well as light intensity, the two factors interact.

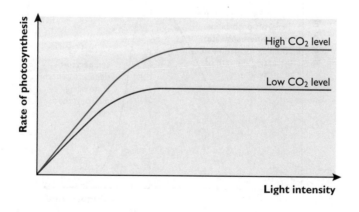

Figure 23 Effect of light intensity on the rate of photosynthesis at two different carbon dioxide levels

Both curves in Figure 23 have the same basic shape, and light intensity eventually becomes non-limiting in both. The rate at which this happens is different because when more carbon dioxide is available, increasing the light intensity allows an even faster reaction.

Finally, consider the graph in Figure 24. It shows the same effects as Figure 23, but at two different temperatures.

Figure 24 Effect of light intensity on the rate of photosynthesis at two different carbon dioxide concentrations at two different temperatures

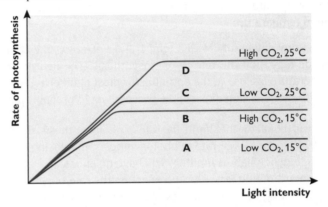

In the region of the graphs where light intensity is non-limiting (horizontal lines), the factors that are limiting are:

- **A** — both temperature and carbon dioxide; increasing either produces an increase in the rate of photosynthesis to level **B** or **C**
- **B** — temperature (the factor that hasn't been increased from **A**); increasing the temperature increases the rate to level **D**
- **C** — carbon dioxide (the factor that hasn't been increased from **A**); increasing the carbon dioxide concentration increases the rate to level **D**

Knowledge check 13

Explain why, in Figure 24, the rate of photosynthesis is higher in curve D than in curve A.

Key facts you must know and understand

Increasing the temperature increases the rate of photosynthesis until it becomes limiting. Figure 25 shows that, after a certain point, increasing the temperature decreases the rate, as the enzymes controlling the reactions start to denature.

Figure 25 Effect of temperature on the rate of photosynthesis

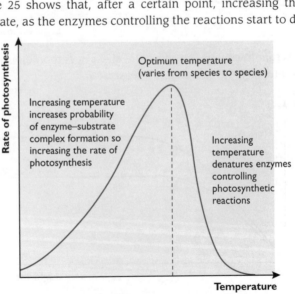

Growing crops in large greenhouses (glasshouses) allows the environment to be controlled to enhance photosynthesis and so increase productivity.

The greenhouse effect, shown in Figure 26, happens in greenhouses as well as in the Earth's atmosphere. Short-wave radiation entering a greenhouse becomes long-wave radiation as it strikes a surface in the greenhouse. Long-wave radiation cannot escape easily through the glass, so the greenhouse warms up.

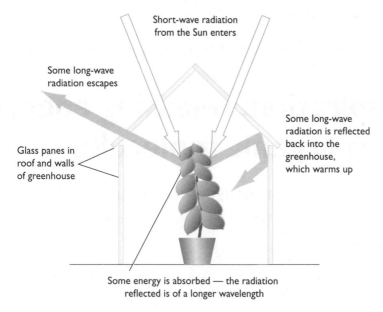

Figure 26 Greenhouse effect

Growers can increase the temperature inside a greenhouse by burning a fossil fuel. This also increases the concentration of carbon dioxide in the air in the greenhouse. Two potentially limiting factors are increased at the same time.

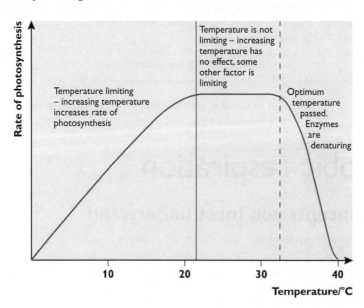

Figure 27 Effect of heating a greenhouse on the rate of photosynthesis

Knowledge check 14

Look at Figure 27. Explain why the owner of this greenhouse should not heat it beyond a temperature of about 20°C.

After studying this topic, you should be able to:

- explain why ATP is a suitable source of energy for biological processes
- describe how ATP and reduced NADP are made in the light-dependent reaction of photosynthesis
- describe how ATP and reduced NADP from the light-dependent reaction of photosynthesis are used in the light-independent reaction

- explain how the structure of a chloroplast is adapted for efficient photosynthesis
- interpret graphs showing the effect of limiting factors on the rate of photosynthesis
- explain and evaluate the ways in which knowledge of limiting factors is applied in commercial greenhouses

Respiration: releasing energy from organic molecules

Key concepts you must understand

Respiration *releases* energy from organic molecules. This energy is used to synthesise ATP.

Examiner tip
Do not to write that respiration *produces* energy. It releases energy and produces ATP.

There are two kinds of respiration:
- **aerobic respiration**, which takes place in the presence of oxygen
- **anaerobic respiration**, which can take place in the absence of oxygen

In both processes, the oxidation of organic molecules is linked to the reduction of coenzymes. The most important of these coenzymes are **NAD**, **FAD** and **coenzyme A**.

Examiner tip
NAD is a coenzyme used in respiration. Do not mix it up with NADP, the coenzyme used in photosynthesis. If it helps, remember that NAD**P** is used in **p**hotosynthesis

Some of the ATP in aerobic respiration and all of the ATP in anaerobic respiration is produced by **substrate-level phosphorylation**. In this process, a compound with a phosphate group attached (XP in the equation below) transfers its phosphate group to ADP, producing a molecule of ATP.

$$\text{Substrate-level phosphorylation}$$
$$\text{XP + ADP} \longrightarrow \text{X + ATP}$$

Most of the ATP produced in aerobic respiration is by **oxidative phosphorylation** and is associated with the transfer of electrons down an electron transfer chain.

Aerobic respiration

Key concepts you must understand

Most of the ATP in aerobic respiration is produced when protons (H^+) diffuse through molecules called **ATP synthase**, located in the inner membranes of mitochondria. As they do so, energy is generated that allows ATP to be produced from ADP and P_i.

Key facts you must know and understand

Most of the ATP-producing reactions of aerobic respiration occur inside mitochondria. Some occur in the matrix of the mitochondria, but most ATP is formed by the ATP synthase in the inner membranes of mitochondria.

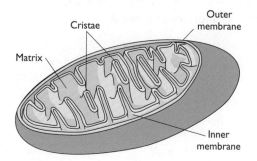

Figure 28 Structure of a mitochondrion

Knowledge check 15

The mitochondrion in Figure 28 is adapted for its function. Suggest how (a) the cristae and (b) the fluid matrix improve the efficiency of aerobic respiration.

Figure 29 shows the four stages in the aerobic respiration of glucose.

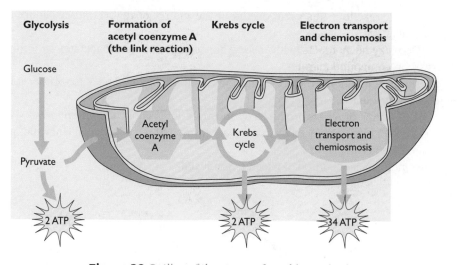

Figure 29 Outline of the stages of aerobic respiration

Glycolysis

Glucose cannot enter a mitochondrion. During glycolysis, glucose is oxidised in the cytoplasm to form **pyruvate**. This molecule can enter a mitochondrion.

During glycolysis:

- two molecules of ATP are used to 'energise' one glucose molecule and 'kick-start' the process
- using phosphate from the two hydrolysed ATP molecules, the glucose molecule is converted into two 3-carbon molecules called triose phosphate
- the two molecules of triose phosphate are oxidised by NAD to form two molecules of pyruvate and reduced NAD

Knowledge check 16

Glucose cannot enter a mitochondrion. Use your knowledge and understanding from Unit 1 to suggest why.

(a) During glycolysis, what is the net gain of ATP per molecule of glucose?

(b) During the oxidation of triose phosphate to pyruvate, ADP is converted to ATP. Where did the additional phosphate come from?

- during this oxidation of two molecules of triose phosphate to two molecules of pyruvate, sufficient energy is released to convert four molecules of ADP into four molecules of ATP

The link reaction

Pyruvate formed in glycolysis enters the matrix of a mitochondrion where it combines with coenzyme A to form **acetylcoenzyme A**.

In this link reaction, each pyruvate molecule loses:
- a carbon atom and a molecule of carbon dioxide is formed
- hydrogen ions (protons), which are taken up by NAD to form one molecule of reduced NAD

Since two molecules of pyruvate are produced per molecule of glucose, for each molecule of glucose the link reaction produces:
- two molecules of acetylcoenzyme A
- two molecules of carbon dioxide
- two molecules of reduced NAD

The Krebs cycle

Like the link reaction, this cycle occurs in the matrix of mitochondria. Figure 30 shows that:
- acetylcoenzyme A reacts with a 4-carbon compound (oxaloacetate) to form a 6-carbon compound (citrate)
- in a series of reactions, the 6-carbon compound is converted back to oxaloacetate
- for each turn of the Krebs cycle,
  - one molecule of ATP is produced by substrate-level phosphorylation
  - four molecules of reduced coenzyme are formed (three of reduced NAD and one of reduced FAD)
  - two molecules of carbon dioxide are produced

Figure 30 The Krebs cycle

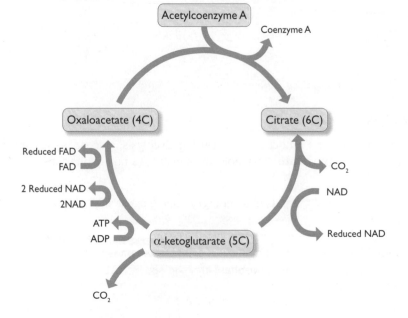

Explain the term 'substrate-level phosphorylation'.

Electron transfer and chemiosmosis

The most important products of the Krebs cycle are reduced NAD and reduced FAD. They are passed to the electron transfer chain where the chemical potential energy they contain is used to produce ATP by oxidative phosphorylation.

The electron transfer chain

The molecules of the electron transfer chain are built into the inner membranes of mitochondria, along with molecules of ATP synthase. Figure 31 shows their arrangement and shows the transfer of electrons along the chain.

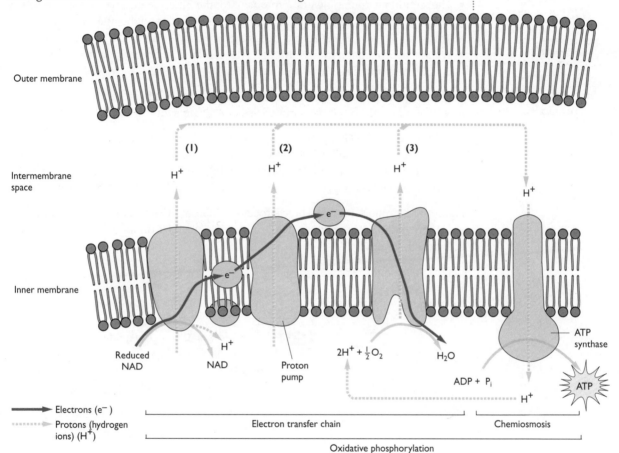

Figure 31 Electron transport and chemiosmosis

- Start at the left of Figure 31. Reduced NAD binds with the first proton pump and releases protons (H⁺) and electrons.
- The electrons are transferred by the proton pump to the next component in the chain.
- Energy released during this transfer of electrons 'powers' the movement of a proton (H⁺) through the pump into the space between the two mitochondrial membranes (the **intermembrane space**).
- As electrons are transferred along the chain, more energy is released, which is used to power the movement of more protons into the intermembrane space through proton pumps 2 and 3.

Knowledge check 19

Use Figure 31 to explain why this method of producing ATP is called oxidative phosphorylation.

- At the end of the chain, the electrons combine with protons and oxygen to form water.
- Since oxygen is the last substance to 'accept' the electrons, it is called the **terminal electron acceptor**.

The chemiosmotic synthesis of ATP

- You can follow in Figure 31 that as the electrons from a reduced NAD molecule pass along the transport chain, proton pumps move three protons (H^+) into the intermembrane space.
- The continual 'pumping' of protons into the intermembrane space creates a diffusion gradient between the intermembrane space and the matrix.
- This gradient results in protons diffusing through ATP synthase molecules.
- When a hydrogen ion moves through ATP synthase, it releases sufficient energy to enable the synthesis of one molecule of ATP.

Examiner tip

You can use the terms hydrogen ion, H^+ and proton interchangeably. They all refer to the same particle.

Electrons from a molecule of reduced FAD only move two hydrogen ions through the proton pumps and so only two molecules of ATP are synthesised.

The formation of ATP by chemiosmosis depends on the electron transfer chain creating the proton gradient. Without oxygen to accept the electrons at the end of the chain, the electron transfer and proton pumping would cease. Because it depends on oxygen, this method of producing ATP is called **oxidative phosphorylation**.

Figure 32 summarises the whole process of aerobic respiration (NADH and FADH are abbreviations for reduced NAD and FAD; Acetyl CoA is an abbreviation for acetylcoenzyme A).

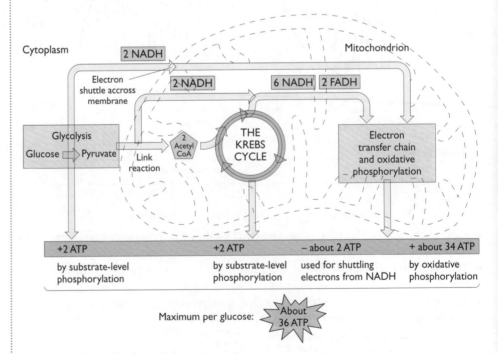

Figure 32 The production of ATP from glucose in aerobic respiration

Anaerobic respiration

Key concepts you must understand

In the absence of oxygen, the electron transfer chain cannot take place because oxygen is not present to act as the terminal electron acceptor. As a result:
- there is a build up of reduced NAD and reduced FAD because they cannot be oxidised
- the link reaction and the Krebs cycle cannot take place

As Figure 33 shows, glycolysis continues because:
- reduced NAD is used to reduce the pyruvate formed in glycolysis to either lactate (animals) or ethanol (plants and yeast)
- in this oxidation-reduction reaction, enough NAD is formed to maintain the reactions of glycolysis

Because only glycolysis (plus the reduction of pyruvate) takes place, anaerobic respiration is much less efficient than aerobic respiration. Each molecule of glucose yields only two molecules of ATP compared with a potential 36 ATP molecules generated by the aerobic pathway.

Key facts you must know and understand

In anaerobic respiration:
- Glucose is converted to pyruvate by glycolysis.
- The net yield of ATP is two molecules per molecule of glucose.
- ATP is only generated by substrate-level phosphorylation during glycolysis.
- The reduced NAD generated in glycolysis cannot pass electrons down the electron transfer chain because there is no oxygen to act as the terminal electron acceptor.
- Reduced NAD is used to reduce pyruvate to either ethanol or lactate. This makes NAD available again and allows the glycolysis reactions to continue.

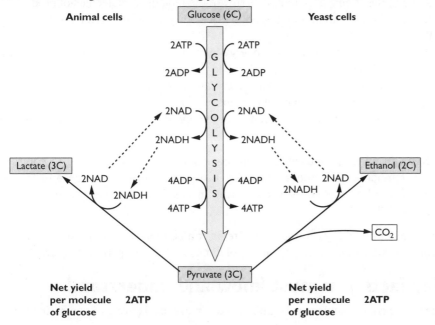

Figure 33 Anaerobic respiration

Knowledge check 20

Use Figure 33 to explain the statement 'The net yield of ATP is two molecules per molecule of glucose'.

Summary

After studying this topic, you should be able to:

- explain how mitochondria are adapted for efficient aerobic respiration
- describe how ATP is synthesised by substrate-level phosphorylation during glycolysis and by oxidative phosphorylation during the electron transfer chain

- explain the role of the link reaction
- describe the role of oxidation–reduction reactions involving coenzymes in aerobic and anaerobic respiration
- explain the importance of oxygen in aerobic respiration

The transfer of energy between organisms

The transfer of energy through ecosystems

Key concepts you must understand

Energy cannot be created or destroyed. It can only be transferred and, in the transfer, converted from one form to another. This is called the **conservation of energy**.

The amount of usable energy in an ecosystem decreases as energy is transferred through it.

- Energy enters ecosystems as light and is transduced to chemical potential energy in the biological molecules within producers as a result of photosynthesis.
- Some of the chemical potential energy in the biological molecules of producers is released in respiration and is used to drive other processes in the producer and is then lost as heat.
- Some of the chemical potential energy in the biological molecules of producers is transferred to herbivorous animals when they feed.
- The same is true when carnivorous animals eat other animals and when decomposers break down dead remains.
- Animals release the chemical potential energy from their food and use it to drive other processes, losing some as heat.

Energy can only be used once by an organism. Energy that is used to contract a muscle cannot be used later in the active transport of ions. This is because in using the energy, it becomes transduced to heat. In this form, the energy cannot be used to create ATP.

The transfer of energy through ecosystems can be represented in food chains, food webs, pyramids of numbers, pyramids of biomass and energy flow diagrams.

Key facts you must know and understand

The biological molecules an organism takes in are used in one of two main ways.

- They are respired to produce ATP, which is used to drive processes such as muscle contraction and active transport; some of the energy is lost as heat.
- They are assimilated into the structure of the organism; the energy in the molecules remains within the organism.

Energy is transferred from one trophic level to another during feeding. This is represented in a **food chain** — for example:

Grass   →   Gazelle   →   Cheetah

Feeding level (trophic level)   *Producer*   *Primary consumer*   *Secondary consumer*

The processes involved in the transfer of energy through this food chain are as follows:

- Light energy enters the grass (the **producer**).
- In photosynthesis, some of the light energy striking a plant is transduced to chemical energy in organic molecules, such as glucose.
- Some of these chemicals are used in growth to build new cells (assimilation) and are retained within the grass.
- Some are used in respiration, during which some of the energy they contain is lost as heat.
- The gazelle (**primary consumer**) eats, digests and absorbs some of the molecules within the grass, transferring chemical energy to its own cells.
- Some of these chemicals are used in growth to build new cells (assimilation) and are retained within the gazelle.
- Some are used in respiration, during which some of the energy they contain is lost as heat.
- The cheetah (**secondary consumer**) eats and digests parts of the gazelle, transferring chemical energy to its cells in the organic molecules it absorbs.
- Some of these chemicals are used in growth to build new cells (assimilation) and are retained within the cheetah.
- Some are used in respiration, during which some of the energy they contain is lost as heat.
- At each stage in the food chain, organisms that are not eaten eventually die and chemical energy contained in their cells is transferred during decay to **decomposers** in the molecules they absorb from the dead organisms.
- Some of these chemicals are used to build new cells and are assimilated into the decomposers.
- Some are used in respiration, during which some of the energy they contain is lost as heat.
- The different feeding levels (producer, primary consumer, secondary consumer and decomposer) are called **trophic levels**.

The flow of energy through this simple food chain is represented in Figure 34.

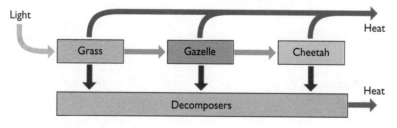

> **Knowledge check 21**
> Suggest **three** reasons why only some of the light energy striking a producer is transduced to chemical energy in organic molecules.

Figure 34 Flow of energy through a food chain

Food chains almost never exist in isolation, but are interlinked to form food webs, such as the ones shown in Figure 35. Changes in any population in a food web can influence the other populations.

- In the soil food web in Figure 35a, a decrease in the number of fungi could lead to an increase in the number of protozoa as there would be more organic matter, which could lead to an increase in the numbers of bacteria, which could lead to an increase in the number of protozoa.
- In the marine food web in Figure 35(b), a decrease in the number of crabeater seals could lead to an increase in the number of petrels, which would be able to eat more fish.

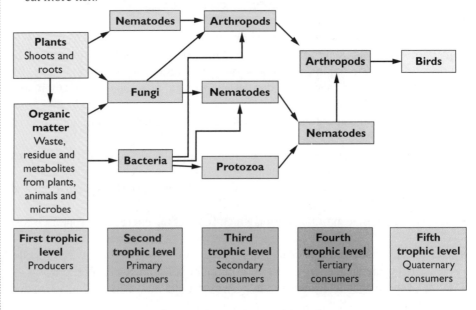

Figure 35 (a) A food web in soil

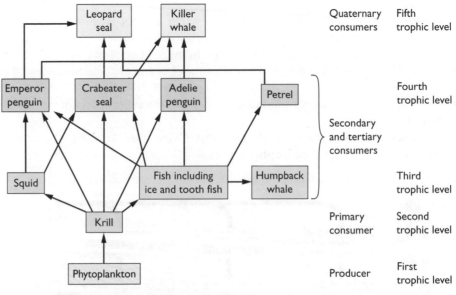

Figure 35 (b) A marine food web

The flow of energy through different ecosystems can be measured. Figure 36 shows one such example.

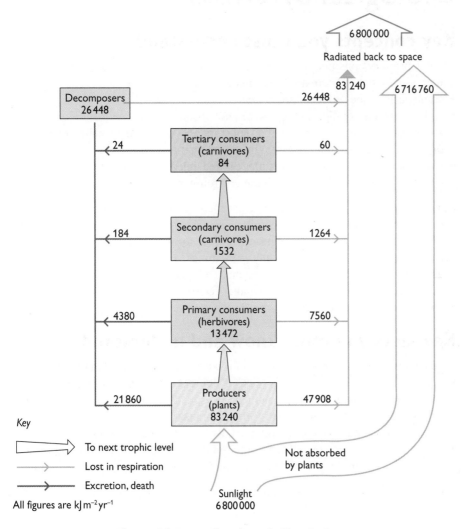

Figure 36 Energy flow through Silver Springs

The number of trophic levels in an ecosystem does not often exceed five because:

- only about 5% of the light shining on plants is used in photosynthesis
- only about 10% of the energy in a trophic level is passed to the next level
- there is too little energy in the fifth trophic level to support another trophic level

Notice in Figure 36 that energy is conserved in the ecosystem as a whole. All the energy entering as sunlight is eventually radiated back to space. Most is reflected off plants, but some passes through the organisms in the ecosystem before it is re-radiated.

Ecological pyramids

Key concepts you must understand

When a primary consumer eats a producer, not all of the materials in the producer end up as consumer. There are losses because:

- some parts of the producer are not eaten
- some parts are not digested and so are not absorbed (they are lost in faeces)
- metabolism of some of the absorbed substances leads to the formation of excretory products, which are released into the environment
- many of the absorbed substances are respired, releasing energy: the carbon dioxide formed at the same time is lost to the environment

Only a small fraction of the substances in the producer becomes incorporated into new cells in a primary consumer. Similar losses are repeated at each trophic level in the food chain, so a smaller biomass is available for growth at successive levels. We can represent these losses graphically as pyramids. The loss in biomass at each level is usually associated with smaller numbers. The exceptions to this occur when:

- very large producers are eaten by smaller consumers
- an organism is parasitised

Key facts you must know and understand

Food chains can be represented as any of the following ecological pyramids:

- A pyramid of numbers represents the total number of the organisms at each trophic level in a food chain, at a given moment, irrespective of biomass (size).
- A pyramid of biomass represents the total biomass of the organisms at each trophic level in a food chain, at a given moment, irrespective of numbers.
- A pyramid of energy represents the amount of energy present in each trophic level of a food chain, irrespective of numbers and biomass, in a given period of time.

The table shows pyramids of numbers, biomass and energy for three different food chains.

Notice that the pyramids of biomass and of energy have a true pyramid shape. No matter what food chain we consider, energy is always lost between trophic levels, so that the energy and the biomass it can support is always less at each trophic level than the one before it.

The pyramids of numbers do not all have this shape, however:

- In the food chain, grasses → grasshoppers → frogs → birds, the individual organisms increase in mass along the chain, so a decrease in biomass necessarily also means a decrease in numbers.
- In the food chain, oak tree → aphids → ladybirds → birds, a single oak tree supports a large number of consumers. The biomass of the single oak tree is, however, massive compared with the individual consumers it supports.
- The food chain oak tree → aphids → ladybirds → birds → mites shows how a pyramid of numbers can be further complicated if the top carnivore is parasitised. Here, each bird would have many tiny mites on its body.

Knowledge check 23

Some of the food eaten by a primary consumer is not digested and absorbed. Suggest **one** carbohydrate that is unlikely to be fully digested.

Food chain	Pyramid of numbers	Pyramid of biomass	Pyramid of energy
Birds ↑ Frogs ↑ Grasshoppers ↑ Grasses			
Birds ↑ Ladybirds ↑ Aphids ↑ Oak tree			
Mites ↑ Birds ↑ Ladybirds ↑ Aphids ↑ Oak tree			

After studying this topic, you should be able to:

- explain that photosynthesis is the main way in which energy enters ecosystems
- explain that energy is transferred through the trophic levels within ecosystems and is eventually radiated back into space
- make quantitative statements about the efficiency of energy transfer between trophic levels
- interpret, and construct from given data, pyramids of number, biomass and energy

Summary

Energy and food production

Key concepts you must understand

Productivity is the energy entering a trophic level that remains as energy in biomass:

- in a given period of time (often a year)
- for a given area of the ecosystem (often a square metre)

It may be expressed in units of mass (e.g. kilograms) or of energy (e.g. kilojoules).

The productivity of producers (plants) is called **primary productivity**.

Not all the biomass produced by plants remains at the end of the year as biomass. Some is respired, and the products of respiration are lost, taking with them some of the energy that was fixed in the biomass.

- **Gross primary productivity** (GP) is all the biomass produced by the plant per m^2 per year.

- **Net primary productivity** (NP) is the biomass that is left per m² per year, after losses in respiration (R) are taken into account. This is the biomass of the plants that will pass either to the primary consumers or to the decomposers following death of the plants.

We can express this relationship in a formula: **NP = GP – R**

The productivity of animals is called **secondary productivity**. As with primary productivity, it is measured in units of mass or of energy per unit of time (usually a year) per unit of area (usually a square metre). The biomass remaining in the animals needs to take into account:

- the amount of energy in ingested food (I)
- the amount of energy lost in respiration (R)
- the amount of energy lost in urine (U)
- the amount of energy lost in faeces (F)

The values of gross and net productivities for the Silver Springs ecosystem in Figure 36 are shown in Figure 37.

Knowledge check 24

Write a formula to show how you would find the net productivity of a consumer.

Knowledge check 25

In Figure 37, decomposers get more of the energy in the herbivore trophic level than do the carnivores. Explain why.

Notice that all the energy in the net productivity passes either to the next trophic level or to the decomposers

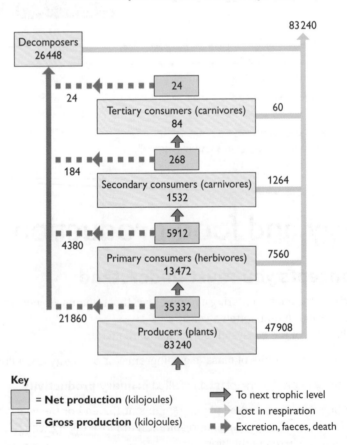

Figure 37 Gross and net productivity in the Silver Springs ecosystem

Key facts you must know and understand

A farmer aims to get as much biomass production for as little input as possible. To do this, a farmer can adopt any of the following practices.

Monoculture

This means growing a single type of crop over a wide area. This practice:
- allows easier control of pests — the number of different pests is often limited by the crop and so specific pesticides can be used
- reduces plant competition for nutrients, space and solar radiation — only the crop plant is taking nutrients from the soil and absorbing light energy
- maximises profit by growing crops with a high gross margin

The use of fertilisers

Harvesting crop plants breaks the natural cycle of decay so that mineral ions are not returned to the soil, which becomes mineral-deficient. Fertilisers add mineral ions to the soil (see Figure 38).

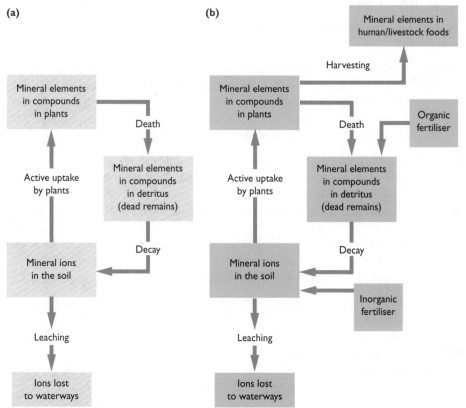

Figure 38 (a) The natural cycle (b) The effects of harvesting and the use of fertilisers

Organic fertilisers are materials produced directly from animals, plants and other living organisms and must be decomposed to release mineral ions. They are, therefore, **slow-release** fertilisers. They include farmyard manure, seaweed, dried blood, sewage sludge and poultry manure.

Knowledge check 26

Give **two** advantages of using farmyard manure rather than inorganic fertiliser.

Inorganic fertilisers do not need to be broken down because they are already in the form of soluble mineral ions. They are, therefore, **quick-release** fertilisers.

Some properties of organic and inorganic fertilisers are compared in the table below.

Property	Organic fertilisers	Inorganic fertilisers
Release of ions	Slow — they must decay	Fast — they are ionic
Solubility	Low	High
Time of application	Early — applied before the crop grows to allow time for decay	Late — applied as the crops approach peak growth as the ions can be absorbed immediately
Contribution to soil structure	Aid water retention and soil crumb structure	None
Consequences of overuse	Few problems because of low solubility and slow release of ions	Eutrophication — high solubility means ions are leached easily into waterways

The use of pesticides

Pests reduce the productivity of crops. Pests include animals, plants (weeds) and some fungi.

Weeds compete with crop plants for the available light, water, carbon dioxide and mineral ions. This is an example of interspecific competition (see pages 10–12). They often out-compete the crop plants because they have higher growth rates than the crops. They establish their root and shoot systems more quickly and so reduce the availability of resources to the crop plants, thus reducing crop yields.

Insect pests can reduce the yield of the crop in a number of ways.
- They can feed on the leaves; this reduces the leaf area and therefore the capacity of the plant for photosynthesis.
- They can feed on the roots, lowering the uptake of mineral ions essential for growth.
- They can feed on sap from the phloem and so disrupt the transfer of sugars manufactured in photosynthesis to other organs.
- They can spread organisms that cause disease.

A **pesticide** is a chemical that helps to control a pest population. Pesticides can be classified according to the type of organism they control, for example:
- **insecticides** kill insects
- **herbicides** kill plants (they are weed killers)
- **fungicides** kill fungi
- **molluscicides** kill molluscs (slugs and snails)

Some pesticides have effects on organisms in the environment other than the pests they are used to control. They might:
- kill useful insects as well as the targeted harmful insects
- persist in the environment for many years before they are finally broken down; they may be taken up by crop plants and so enter humans through the food chain
- accumulate along food chains (**bioaccumulation**), e.g. DDT

Biological control

This involves introducing a natural parasite or predator of the pest into the area. The aim is to reduce the pest population to a level that does not cause major damage.

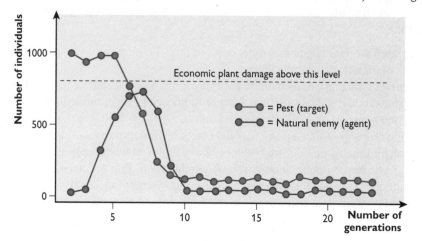

Figure 39 Biological control

Knowledge check 27

Figure 39 shows that a biological control agent reduces the number of pests but does not eradicate them. Explain why it does not eradicate them.

The following are examples of biological control methods:
- Introduce a **predator** — e.g. ladybirds have been used to control aphid populations in orange groves.
- Introduce a **herbivore** — e.g. a moth native to South America was introduced to Australia to control the 'prickly pear' cactus.
- Introduce a **parasite** — e.g. larvae of the wasp *Encarsia* are introduced into greenhouses to control whiteflies on tomato crops.
- Introduce **sterile males** — this reduces the number of successful matings and so reduces the pest numbers.
- Use **pheromones** — these animal sex hormones are used to attract the males or females, which are then destroyed, e.g. male-attracting pheromones are used to control the damson-hop aphid, reducing damage to plum crops.

Biological control has several advantages over the use of pesticides:
- Pests do not usually develop resistance to a predator or parasite.
- Biological control agents are usually much more specific than pesticides; for example, a carefully chosen predator will target only the pest, whereas a pesticide might target all the animals of a particular group (an insecticide might kill many kinds of insect).
- Once a natural predator or parasite has been introduced, no further reintroductions are necessary, whereas pesticides must be re-applied regularly.

Biological control is not always appropriate since:
- the proposed control agent might affect populations other than the pest
- the control agent might not be able to reproduce in the new environment
- reducing the numbers of one specific pest might allow another pest to fill the original pest's ecological niche
- it is not an appropriate method for controlling pests of stored grain. The grain would become contaminated with the dead bodies of both pest and control agent.

Examiner tip

If you are asked to make a comparison, you must compare like with like. An answer such as 'A biological control agent is usually specific to only one type of pest whereas a pesticide usually kills far more species than just the pest' would gain credit. An answer such as 'A biological control agent is specific but pesticides might build up in food chains' would not gain credit.

Integrated control systems

Often, neither chemical control nor biological control alone is really effective in controlling pests. Biological control is often enhanced when low levels of pesticides are used at the same time. This is a simple example of an **integrated control system**.

In integrated crop management, most or all of the following would be considered as methods of maximising productivity:

- selecting crops that are adapted to the type of soil and the climate in the area
- selecting crops that have some resistance to known pests in the area
- choosing appropriate methods of pest control
- rotating crops grown in a particular field so that the same pests do not build up in the soil and the same ions are not continually removed by the crop
- using fertilisers (organic, inorganic or a combination) that are appropriate to the conditions, to replace the mineral ions removed by cropping
- appropriate treatment and storage of the final crop (to minimise damage by pests)
- irrigation of the soil (where necessary)

By using a combination of these techniques, a farmer can reduce damage by pests and ensure that crops have a continuous supply of mineral ions.

Intensive rearing of livestock

Traditionally, stock animals were allowed to wander within the boundaries of fields or pens. The main principles of intensive farming practices include:

- feeding a precisely controlled diet that ensures less of the food is lost in faeces
- restricting the movement of the animals so that less energy is lost this way and more energy is used in growth
- keeping the animals in a warm environment so that less energy is lost as heat to the environment and more is used in growth
- using hormone injections or supplements to increase the rate of growth
- using antibiotic injections to increase the rate of growth

Knowledge check 28

Link the first three items in the list of intensive farming practices to components of the equation: net productivity = energy ingested − energy lost in respiration − energy lost in faeces ($NP = I - R - F$).

Summary

After studying this topic, you should be able to:

- use the equation: net productivity = gross productivity − energy losses
- identify the energy losses of plants and animals in the above equation
- explain ways in which productivity is increased by farming practices that increase the efficiency of energy conversion
- use your understanding of farming practices that increase productivity to evaluate their associated economic and environmental issues and to consider the ethical issues they raise

How elements are cycled in ecosystems

The carbon and nitrogen cycles

Key concepts you must understand

Life on Earth is carbon-based; carbohydrates, lipids, proteins, nucleic acids and ATP all contain carbon.

Proteins, nucleic acids and ATP contain nitrogen as well as carbon.

There is a fixed amount of carbon and nitrogen within an ecosystem; normally these elements are recycled.

Microorganisms in the soil decompose dead organic matter. As by-products of the decay process, carbon atoms and nitrogen atoms in organic molecules are converted into forms that can be taken up and used by plants. In this way, these elements are recycled.

Key facts you must know and understand
The carbon cycle

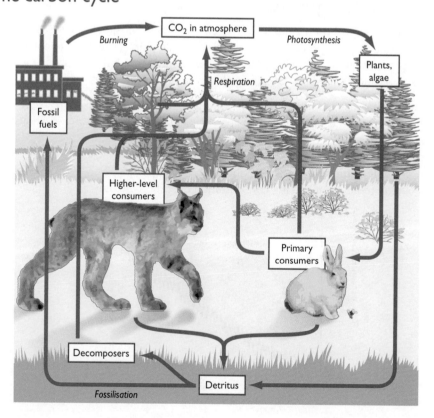

Figure 40 The carbon cycle

Examiner tip
You will not gain credit for writing that *plants take up carbon* or that *plants take up nitrogen*. You must refer to carbon dioxide and to nitrates.

Knowledge check 29
Name three items you would expect to be in the box labelled detritus in Figure 40.

The main processes involved in cycling the element carbon through ecosystems are:

- photosynthesis — fixes carbon atoms from carbon dioxide into organic compounds, such as glucose
- feeding and assimilation — feeding passes carbon atoms in organic molecules to the next trophic level in the food chain where they are assimilated into (become part of) the body of that organism
- respiration — releases carbon dioxide from organic compounds
- fossilisation — sometimes dead material does not decay fully due to the conditions in the soil, and fossil fuels (e.g. coal, oil and peat) are formed
- combustion — fossil fuels are burned, releasing carbon dioxide into the atmosphere

Figure 40 summarises the carbon cycle.

Too much carbon dioxide: global warming

The concentration of carbon dioxide in the atmosphere is affected by:

- the removal of carbon dioxide from the air by plants for use in photosynthesis
- the addition of carbon dioxide to the air by all organisms as a result of respiration
- the addition of carbon dioxide to the air as a result of volcanic eruptions and natural fires

The levels of carbon dioxide in the atmosphere fluctuate as overall rates of photosynthesis and respiration change. In winter, the amount of photosynthesis is reduced due to:

- cooler temperatures
- shorter day length
- loss of leaves by many plants

As a result, less carbon dioxide is absorbed from the atmosphere. Respiration may also be reduced, but it produces more carbon dioxide than is used in photosynthesis. The concentration of carbon dioxide rises. In summer, the balance is reversed and the concentration of carbon dioxide in the atmosphere falls.

The recent increase in the amount of carbon dioxide in the atmosphere is probably due to the increased combustion of fossil fuels by humans. Figure 41 shows the measurements of the concentration of carbon dioxide at Mauna Loa on Hawaii, which demonstrate:

- the annual fluctuations described above
- an overall trend of increasing concentration of carbon dioxide

Figure 41 Carbon dioxide concentration in the air at Mauna Loa

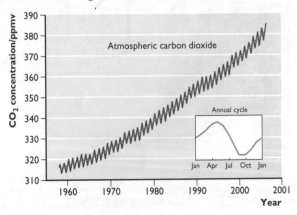

Knowledge check 30

Explain the fluctuations shown in the curve in Figure 41.

Carbon dioxide is a greenhouse gas. Figure 42 shows how carbon dioxide may be a factor in global warming by contributing to the greenhouse effect.

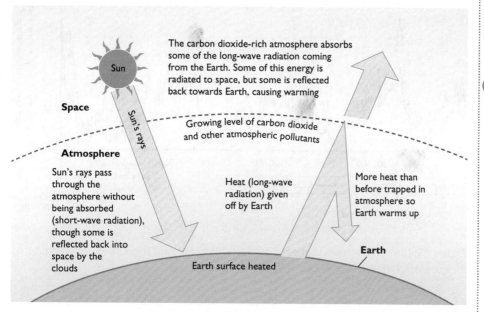

Figure 42 The greenhouse effect

However, the evidence is correlational so no definite cause and effect can be proved. The evidence is also conflicting.

Over the past 30 years, the carbon dioxide concentration of the atmosphere has increased and so has the temperature of the Earth. If we look over longer periods of time, however, the correlation is not so clear-cut. As Figure 43 shows, the changes in temperature and carbon dioxide concentration since 1850 (when the industrial revolution began) do not always coincide.

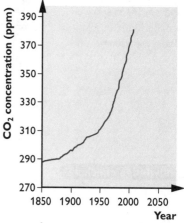

Figure 43 (a) Changes in carbon dioxide concentration since 1850

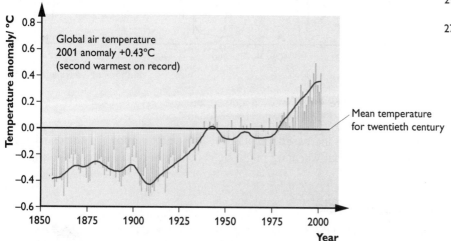

Figure 43 (b) Changes in temperature since 1850

Examiner tip

You might be asked to evaluate a conclusion, such as carbon dioxide causes global warming. Try to use the evidence in the question to give a balanced evaluation. In other words, give at least one way that the evidence supports the conclusion and one way that it does not support the conclusion.

Knowledge check 31

Explain how the data in Figure 43 can be used to conclude that rising CO_2 concentrations might not have caused global warming.

Figure 44 shows that over a longer period than this, there does seem to be a correlation between carbon dioxide concentration and temperature.

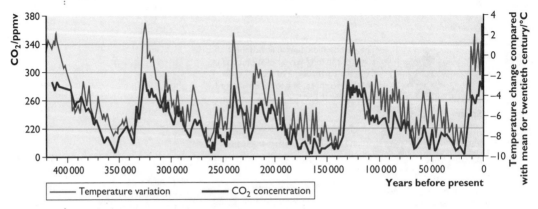

Figure 44 Changes in temperature and carbon dioxide concentration over the past 400 000 years

Examiner tip

Judging by their examination answers, many candidates' knowledge of global warming seems to be restricted to polar bears and ice. Try to show a broader understanding of global warming than this.

Figure 45, however, shows that over an even longer period there does not seem to be a clear correlation.

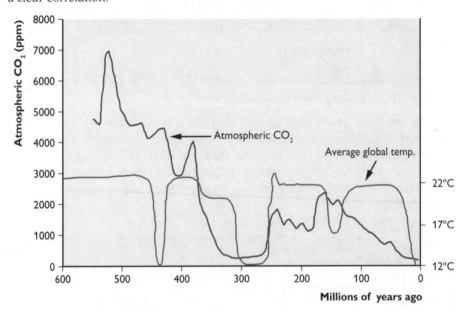

Figure 45 Changes in temperature and carbon dioxide concentrations over 600 million years

Knowledge check 32

Figure 44 could be used to show that increases in CO_2 concentration cause global warming and that global warming causes an increase in CO_2 concentrations. Explain how both conclusions could be justified from the same data.

Bear in mind also that increased temperature will cause carbon dioxide dissolved in water to escape. Global warming might be causing the increase in carbon dioxide concentration.

The nitrogen cycle

The main processes in the nitrogen cycle on land are listed below.

- Plants absorb nitrates from the soil and use the nitrates to form amino acids, which they use to synthesise proteins.

- Plants are eaten by animals, which digest the proteins into amino acids, absorb the amino acids and assimilate them into animal proteins.
- Plants and animals die, leaving a collection of dead materials (detritus), which contain the nitrogen still fixed in organic molecules.
- Decomposers (**saprobionts**) hydrolyse molecules in the dead organisms (and the organic molecules in urine and faeces), releasing ammonium ions (NH_4^+) into the soil.
- **Nitrifying bacteria** oxidise the ammonium ions first to nitrites (NO_2^-), then to nitrates (NO_3^-), which are taken up by the plants.

In addition to these processes, **nitrogen-fixing bacteria**, living free in the soil and in nodules on the roots of legumes (plants with 'pods' such as peas, beans, lentils and clover), 'fix' nitrogen gas into molecules of ammonia. This ultimately increases the nitrates available to plants.

Denitrifying bacteria reduce nitrate to nitrogen gas that escapes from the soil. This decreases the nitrates available to plants.

Figure 46 summarises the nitrogen cycle.

> **Knowledge check 33**
>
> Name **one** other nitrogen-containing compound that plants will make using the nitrates they absorb.

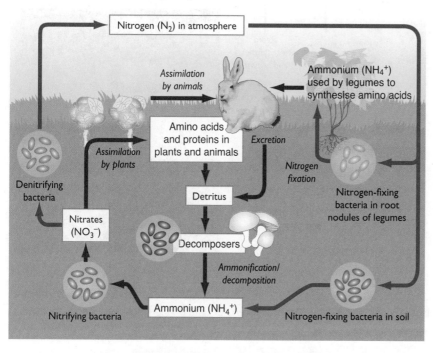

Figure 46 The nitrogen cycle

Too much nitrogen: eutrophication

The soluble nitrate ions in inorganic fertilisers can be carried into nearby waterways. This is called **leaching**.

If inorganic fertilisers are overused, too many nitrate ions are leached and the following series of events can happen:
- Algae in the waterway use the additional nitrate ions to synthesise more proteins, so their populations grow rapidly.

- The increased algal growth forms a mat over the water surface (if the algae are filamentous) or algal bloom (if the algae are unicellular).
- The algal mat or bloom reduces the transmission of light to lower levels of the waterway.
- Rooted plants growing at these lower levels cannot photosynthesise and so they die.
- Algae also start to die as they use up the mineral ions.
- Microorganisms decompose the dead plants and algae and reproduce rapidly as more and more algae and plants die.
- The aerobic respiration of these microorganisms uses increasing amounts of oxygen as their numbers increase.
- The concentration of oxygen in the water falls dramatically and many animals die.

This whole process, summarised in Figure 47, is known as **eutrophication**.

Knowledge check 34

Suggest one reason why eutrophication is (a) more likely to occur in hot weather, (b) less likely to occur in moving water than in still water.

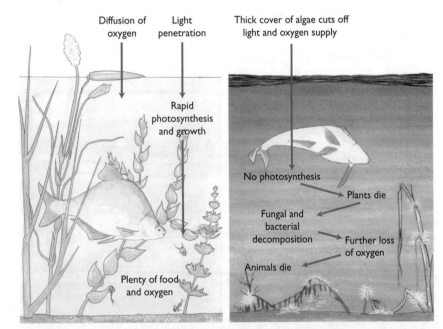

Figure 47 Effects of eutrophication

Summary

After studying this topic, you should be able to:

- describe the carbon and nitrogen cycles, including the role of microorganisms
- explain the role of greenhouse gases, such as carbon dioxide, in bringing about global warming
- evaluate data linking global warming and carbon dioxide concentration
- interpret data linking global warming with the distribution of animal and plant populations, including crop plants
- explain, and interpret data relating to, eutrophication

Succession: how ecosystems change over time

Succession

Key concepts you must understand

Organisms that survive in an environment are adapted to that environment. If the environment changes, those organisms will be less well adapted to the changed environment. Other organisms may be better adapted to the changed environment and out-compete the original organisms, which may become locally extinct.

The ecosystems that exist today did not always exist. They have developed from other previous ecosystems by **succession**.

Succession often involves an initial colonisation of a hostile environment; the colonisers cause changes in the abiotic environment. These changes allow other species to enter the area because it is now less hostile. In turn, these new species further modify the environment, making it less suitable for the colonisers and more suitable for yet newer species. As a result, further changes in the community occur. Therefore, the complexity of the food webs, which make up the community of the developing ecosystem, increases. The succession leads to a final, complex state called the **climax community**.

Key facts you must know and understand

An area with virtually no organisms may be colonised by a species with appropriate adaptations. As the population of this **pioneer species** grows, it interacts with the environment and changes it. The changes may allow other species to enter the area and cause further changes.

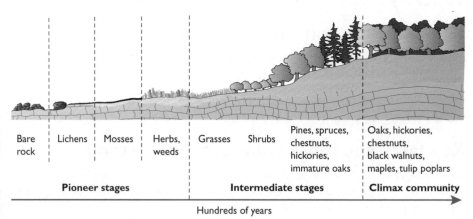

| Bare rock | Lichens | Mosses | Herbs, weeds | Grasses | Shrubs | Pines, spruces, chestnuts, hickories, immature oaks | Oaks, hickories, chestnuts, black walnuts, maples, tulip poplars |

Pioneer stages   **Intermediate stages**   **Climax community**

Hundreds of years

Figure 48 An example of ecological succession

This type of succession is illustrated in Figure 48 and includes the following stages:

- Lichens are the pioneer species that can grow on bare rock.
- The growth of the lichens on the rock, together with weathering, breaks up the rock surface and dead lichens add organic matter; a soil begins to form.
- Mosses start to grow in the developing soil and out-compete the lichens; the mosses add more organic matter to the soil, making it more fertile.
- This process of changing the environment and making it less hostile is repeated as grasses, shrubs and finally trees colonise the area.
- As more and bigger plants colonise the area, they offer more habitats and niches to animals and so the species richness increases.
- The mixed hardwoods (oak, hickory, chestnuts etc.) form the climax community; the ecosystem will not become any more complex.

As different types of vegetation enter the area, they affect the amount and depth of soil. This, in turn, allows other types of plant to enter. The increasing complexity of the plant community creates more and more ecological niches and so more animals will enter the area. The species diversity increases through the succession, until a climax is reached.

Forest in Europe does not become as complex as tropical rain forest because of the climate. Our mixed forests are said to be a **climatic climax** community.

Grassland in much of Europe would revert to woodland/forest if sheep and cattle did not graze it. They nip off the growing points at the tips of young tree shoots. Grasses grow from ground level and so can re-grow. These grasslands are a **grazing climax**.

Where a succession starts from bare, previously uncolonised, ground or from a newly formed pond with no life, the succession is a **primary succession**. Sometimes, communities are destroyed, e.g. by fire. When a new succession begins in such an area it is a **secondary succession**.

Succession and conservation of ecosystems

Key concepts you must understand

Succession leads to change in an ecosystem, whereas conservation might require the ecosystem to be maintained in more or less the same condition.

People conserving ecosystems must interfere with, and manage, the process of succession in such a way that the ecosystem does not change significantly.

Key facts you must know and understand

The management of heather moorland is used here as just one example of managing succession. You should use it to understand the principles that you could apply to another example given in an examination.

Heather moorland is often maintained as a habitat for grouse, but it also provides a habitat for other animals that would not exist elsewhere.

Knowledge check 35

During the succession shown in Figure 48, shrubs out-compete grasses and replace them. For what are these plants likely to be competing?

Examiner tip

Examiners might ask you to 'think out of the box' in data questions testing your understanding of ecological succession. Make sure you can relate data such as the effect of light intensity on the rates of photosynthesis of competing plants to the stages of succession.

Heather plants are small woody shrubs. Where there is a dense cover of heather, it provides an ideal habitat for grouse. As the heather grows taller, the cover becomes less dense and the habitat is less suitable for grouse. There are four main stages in the life cycle of heather plants, described in the table below.

Phase	Duration/years	Productivity and biomass of heather plants	Appearance of heather plants
Pioneer	0–6	Low biomass, high productivity	Small separate shrubs
Building	6–15	High biomass, high productivity	Individual plants intertwine to form a dense canopy
Mature	12–28	High biomass, decreasing productivity	Gaps begin to appear in the centre
Degenerate	20–30	High biomass, low productivity	Gaps increase and other plants begin to grow in the gaps

The degenerate phase is the start of a succession to mixed woodland as trees begin to grow in the gaps in the centre of heather plants.

This is prevented by burning the heather at regular intervals. The fire is carefully managed so that it does not become too intense and damage the underground parts of the plants. These then re-grow and the cycle starts again. Succession is prevented and the heather moor is conserved.

Knowledge check 36

Use the table to suggest a suitable interval at which heather should be burned. Explain your answer.

Summary

After studying this topic, you should be able to:
- describe and explain succession from a pioneer community to a climax community
- explain the changes in species diversity that occur during succession
- explain that conservation frequently involves management of succession
- use your understanding of succession and conservation to evaluate conflicting evidence relating to the conservation of species or habitats

Inheritance: passing on genes from one generation to the next

Key concepts you must understand

This topic contains few new facts for you to learn. Instead, it uses concepts you learnt in Unit 2 in new ways.

You learnt in Unit 2 that:
- the chromosomes in most animal and plant cells (except the sex cells) exist in **homologous pairs**. The total number of chromosomes in such cells is the **diploid number** of chromosomes
- as a result of **meiosis**, **gametes** (sex cells) have only one chromosome from each homologous pair. Gametes have the **haploid** number of chromosomes (half the diploid number)

- fertilisation is a random event; any male sex cell could fertilise any female sex cell
- genes controlling the same feature are present at the same place (**locus**) on homologous chromosomes
- **alleles** are different versions of the same gene. Figure 49 shows that, although homologous chromosomes carry genes for the same features in the same sequence, the alleles of those genes might be different
- each physical, chemical and serological feature (**phenotype**) within a species results from an interaction of genetic factors (**genotype**) for that feature and environmental factors

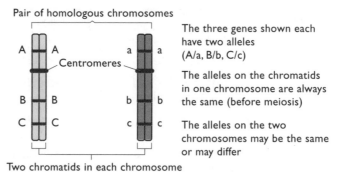

Figure 49 Homologous chromosomes

Let's take an example of one feature, the ability of humans to taste a substance called phenylthiocarbamide (PTC). This feature is controlled by a gene with two alleles: one allele enables people to taste PTC and the other allele does not.

Since body cells are diploid, each body cell contains two chromosomes with a gene for the ability to taste PTC. The alleles of this gene might be:

- the same (e.g. two alleles for tasting PTC); this genotype is **homozygous** for the feature
- different (e.g. one allele for tasting PTC and one non-tasting allele); this genotype is **heterozygous** for the feature

Knowledge check 37

If the allele coding the ability to taste PTC is represented by **T** and the other allele is represented by **t**, what is the genotype of (a) a heterozygous individual and (b) the gametes produced by this individual?

Finally, alleles may be **dominant**, **recessive** or **codominant**. The different types of allele produce different effects on the phenotype under different circumstances, as shown in the table below.

Nature of allele	Effect shown in phenotype of homozygous individual?	Effect shown in phenotype of heterozygous individual?
Dominant	Yes	Yes
Recessive	Yes	No
Codominant	Yes	Yes — both codominant alleles produce their effects

Monohybrid inheritance

Monohybrid inheritance involves the inheritance of a feature controlled by a single gene. You need to be able to deal with monohybrid crosses that involve:

- dominant, recessive and codominant alleles
- multiple alleles

Monohybrid inheritance with dominant and recessive alleles

The flowers of pea plants may be purple or white. A single gene controls this feature. The table below shows the possible genotypes produced by the two alleles of this gene and the phenotypes that result from them.

Example 1: flower colour in pea plants

In the table, **P** represents the dominant allele for purple flower colour and **p** represents the recessive allele for white flower colour.

Genotype	Description of genotype	Phenotype	Reason for phenotype
PP	Homozygous for dominant allele (purple)	Purple flowers	Only the dominant allele (purple flowers) is present
Pp	Heterozygous	Purple flowers	The dominant allele (purple flowers) is always expressed in the heterozygote
pp	Homozygous for recessive allele (white)	White flowers	Only the recessive allele (white flowers) is present

Parental phenotype	Purple flowers	White flowers	Plants from pure-breeding lines are cross-pollinated
Parental genotype	PP	pp	Both are homozygous
Parental gametes	(P)	(p)	Gametes are haploid, so contain only one allele from a pair — only one type of gamete from each parent
F1 genotype	Pp		All F1 plants are heterozygous with purple flowers — purple allele is dominant
F1 gametes	(P) (p) (P) (p)		All F1 plants can produce two types of gamete — half with the purple flower allele and half with the white flower allele (two sets are shown to represent male and female gametes)

F2 genotypes

	(P)	(p)
(P)	PP	Pp
(p)	Pp	pp

This is the standard way of showing all the possible fertilisations and possible combinations of alleles in the F2 generation

In this instance, three of the possible four combinations contain at least one dominant allele and so have purple flowers, while only one of the four has two recessive alleles to give white flowers

F2 phenotypes   3 purple flowers : 1 white flower

Figure 50 A cross between pure-breeding (homozygous) pea plants

Breeding pea plants over two generations and observing their flower colours enables us to see some of the patterns in monohybrid inheritance. We would use the following procedure:

- Cross-pollinate pure-breeding (homozygous) purple-flowered plants with pure-breeding (homozygous) white-flowered plants.

Examiner tip
Sometimes when asked to use a genetic diagram, you will be given a blank version of the labelled layout shown in Figure 50. Sometimes you will be given a blank space and asked to draw your own. Practise producing these diagrams and show all the stages — it will increase your confidence when you sit your BIOL4 paper.

- Collect the seeds formed and germinate them to give a new generation of pea plants — the **offspring (1)** generation (sometimes referred to as the **F1** generation).
- *Self*-pollinate the plants of the offspring (1) generation.
- Collect the seeds formed and germinate them to give the offspring (2) generation (sometimes called the **F2** generation).

The genetic diagram in Figure 50 summarises such a cross. We always use this layout when explaining a pattern of inheritance. It is referred to in the AQA specification as a **fully labelled genetic diagram**.

Monohybrid inheritance with codominant alleles

The flower colour of snapdragons is also controlled by a single gene with two alleles. The alleles are, however, codominant. The table below shows the possible genotypes produced by the codominant alleles of this gene and the phenotypes that result from them. Notice that we use a notation to represent codominant alleles that is different from the one we use to represent dominant and recessive alleles.

Example 2: flower colour in snapdragons

In the table, C^R represents the codominant allele for red flower colour and C^W represents the codominant allele for white flower colour.

Genotype	Description of genotype	Phenotype	Reason for phenotype
C^RC^R	Homozygous for codominant red allele	Red flowers	Only the red codominant allele is present
C^RC^W	Heterozygous	Pink flowers	Both codominant alleles are present; both express themselves so red *and* white pigments are produced, resulting in pink flowers
C^WC^W	Homozygous for codominant white allele	White flowers	Only the white codominant allele is present

Figure 51 is a labelled genetic diagram showing a cross between two pink-flowered snapdragon plants. You know from the above table that pink-flowered plants must be heterozygous for this feature, i.e. C^RC^W.

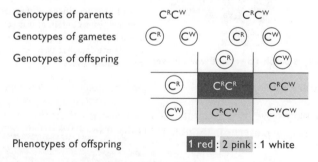

Figure 51 A cross between two pink-flowered snapdragon plants

The ratio of genotypes of the offspring in Figure 51 is the same as that for the F2 genotypes in Figure 50, i.e. homozygous for one allele to heterozygous to homozygous for the other allele = 1:2:1. But, because in Figure 51 we are dealing with codominant alleles, the phenotype ratio is also 1:2:1 (and not 3:1 as we had in Figure 50). As you develop your skills, you will come to recognise the importance of phenotype ratios when interpreting questions in BIOL4 tests.

Monohybrid inheritance with multiple alleles

The two examples of monohybrid inheritance above involved genes with two alleles. We say that a gene has **multiple alleles** if it has more than two alleles. The human ABO blood group, for example, is controlled by a gene with three alleles. Of course, any one person only has two alleles for this blood group — one on each chromosome of a homologous pair.

The ABO blood group of a person is determined by the presence or absence of two antigens (antigen A and antigen B) on the surface of their red blood cells. The three alleles involved are:

- I^A which controls the production of the A antigen.
- I^B which controls the production of the B antigen.
- I^O which results in neither antigen being produced.

Alleles I^A and I^B are codominant; but I^O is recessive to both. The table below shows the possible genotypes produced by the three alleles of this gene and the phenotypes (ABO blood groups) that result from them.

Genotype	Phenotype (ABO blood group)
$I^A I^A$, $I^A I^O$	A
$I^B I^B$, $I^B I^O$	B
$I^A I^B$	AB
$I^O I^O$	O

Figure 52 shows it is possible for a couple to have four children each with a different ABO blood group. Remember, the I^A and I^B alleles are codominant and the I^O is recessive.

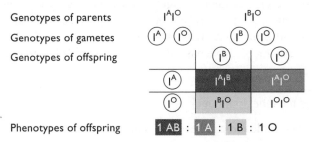

Figure 52 Inheritance of the ABO blood group

Pedigrees

A **pedigree** is a way of representing the inheritance of a feature over several generations in a particular 'family'. In a pedigree:

- each horizontal row represents a generation

Knowledge check 38

What ratio of offspring would you expect if a snapdragon with pink flowers was crossed with a snapdragon with white flowers? Use a labelled genetic diagram to explain your answer.

Knowledge check 39

A student discovered that he had blood group O. He became worried because his mother was blood group A and his father was blood group B. Need he have worried? Explain your answer.

- vertical lines link one generation to the preceding generation
- a female is represented by a circle and a male is represented by a square

Examples of pedigrees

The pedigrees below show:

- the pattern of inheritance of PTC tasting (if you can taste it, it is very bitter indeed)
- the pattern of inheritance of albinism in a family

Both characters are controlled by a single gene with two alleles.

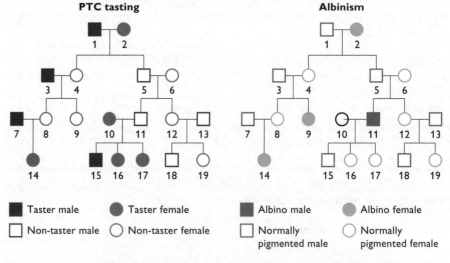

Figure 53 Pedigrees for PTC tasting and albinism

Examiner tip

In an examination, you could be given a pedigree but not be told which feature is determined by the dominant allele. Don't panic; somewhere in the pedigree will be parents with the same feature who produce children with a different feature. Find them and work from there.

To work out what is going on in a pedigree, look for individuals showing the recessive feature. These individuals must be homozygous and so:

- they must have inherited one recessive allele from each parent
- they will pass on one recessive allele to all their children

Look at the PTC pedigree example in Figure 53: individuals 1 and 2 can taste PTC, but their children (individuals 4 and 5) cannot. From this we deduce that:

- individuals 1 and 2 must have the tasting allele
- individuals 1 and 2 must have passed on a non-tasting allele to each of their non-tasting offspring (individuals 4 and 5)
- since individuals 1 and 2 must have both alleles and, because they are tasters, the tasting allele must be dominant

If you had not been told which allele was dominant, you would use similar logic to interpret the albinism pedigree.

- Since individuals 3 and 4 have normally pigmented skin, each must have an allele for normal pigmentation.
- They must have passed on alleles for albinism to produce an albino child (individual 9).
- Individuals 3 and 4 must have both alleles and, because they have normally pigmented skin, the allele for normal pigmentation must be dominant and the allele for albinism must be recessive.

AQA A2 Biology

You would apply exactly the same logic if you had chosen couple 5 and 6 or couple 7 and 8; there was nothing special about choosing couple 3 and 4.

Worked example

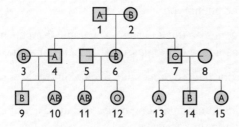

Figure 54 A blood-group pedigree

What are the blood groups of individuals 5 and 8? Explain your answer.

Answer

Individual 5 must pass on an I^O allele to individual 12 who must have two (as she is blood group O). He must also pass on an I^A allele to individual 11, who is blood group AB and cannot have inherited I^A from her mother (who is blood group B).

Individual 5 therefore has the genotype I^AI^O and is blood group A.

Individual 8 must be blood group AB. Individual 7 is blood group O (genotype I^OI^O) and so can only pass on I^O alleles. The I^A and I^B alleles in children 13, 14 and 15 *must* have come from individual 8.

Inheritance of sex in humans

Human sex is determined by the sex chromosomes: the genotype of females is **XX** and that of males is **XY**. Figure 55 shows how the sex of a couple's children is determined.

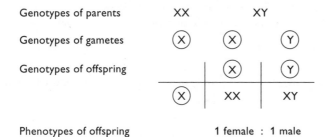

Figure 55 Inheritance of sex in humans

Inheritance of sex-linked characters

A **sex-linked character** is controlled by a gene with a locus on one of the sex chromosomes. Because the Y chromosome is very short, with few loci, most sex-linked characteristics are controlled by genes on the X chromosome.

The X and Y chromosomes are largely non-homologous, i.e. they carry different genes. Therefore:

- in males, a recessive allele on the single X chromosome will show its effect in the phenotype (as there will be no equivalent gene on the Y chromosome)
- in females, a recessive allele must be present on both X chromosomes for it to show its effect in the phenotype

A male (XY) inherits genes on the X chromosome only from his mother. She can only pass on the X chromosome; a father passes on a Y chromosome to his son.

Knowledge check 40

Haemophilia is controlled by an allele of a gene found on the X chromosome. A woman whose blood clots normally has a father who is haemophiliac. What proportion of her eggs would you expect to carry the allele for haemophilia? Explain your answer.

Sex-linked features controlled by recessive alleles on the X chromosome share the following characteristics:

- They are much more common among males (because females must inherit two X chromosomes carrying the recessive allele, whereas males must inherit only one).
- Affected males inherit the sex-linked allele from their mothers, who are often heterozygous and not showing the feature themselves. Females who are heterozygous for the condition are called **carriers**.
- As a result of the above, the feature often 'skips' a generation and then appears *in the males only*.
- Affected females inherit one allele from each parent (so the father will always show the feature as well).

Features determined by recessive alleles carried on the X chromosome include red–green colour blindness and haemophilia.

We represent the alleles of sex-linked characters using the appropriate sex chromosomes as well as the alleles. For red–green colour blindness, **B** represents the dominant allele for normal vision and **b** represents the recessive allele for red–green colour blindness. Putting these alleles on their respective sex chromosome, the possible genotypes and phenotypes are:

- X^BY — male with normal colour vision
- X^bY — male with red–green colour blindness
- X^BX^B — female with normal colour vision
- X^BX^b — carrier female (with normal colour vision)
- X^bX^b — female with red–green colour blindness

Example

The pedigree shows the inheritance of red–green colour blindness in a family.

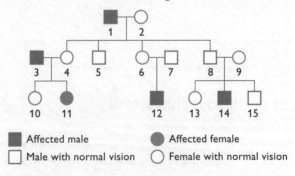

Figure 56 Inheritance of red–green colour blindness

If you were not told that this was a sex-linked feature, there are several clues.
- It clearly skips a generation (e.g. individual 1 via 6 to 12).
- It is more common in the males.
- The only affected female has an affected father.

To work out the genotypes of individuals in a pedigree of a sex-linked feature, begin with a genotype of which you can be certain. This can be an affected male (e.g. genotype X^bY — with the affected X chromosome inherited from his mother), an unaffected male (X^BY) or an affected female (X^bX^b — each parent has passed on one affected X chromosome).

You can now work backwards and forwards from this known starting point. In the example given in Figure 56, what are the genotypes of individuals 9 and 10?
- Individual 9 is the mother of individual 14 — an affected male (X^bY). The X^b chromosome can only have come from the mother (individual 9) who is unaffected. She must therefore have the X^B chromosome and her genotype must be X^BX^b.
- Individual 10 is the daughter of individual 3 and is unaffected. She inherits an X chromosome from each parent and so must inherit X^b from individual 3 (affected male). She must be X^BX^b — a carrier female.

> **Examiner tip**
> When you are solving pedigrees like this, write on the pedigree diagram the genotypes of which you can be certain at the outset. This will help you to see where the various affected chromosomes have come from.

Genes in populations

Key concepts you must understand

A population is a group of individuals of the same species found in a habitat at any given time.

The **gene pool** is the alleles of all the genes that are present in a population.

Populations with a large gene pool have high genetic diversity. Such populations can withstand changes in the environment, as some individuals will have the necessary adaptations to survive.

Populations with a small gene pool have low genetic diversity. Such populations are more likely to become extinct as a result of changes in the environment.

Within a gene pool, every allele or gene variant has a particular **frequency**.

The frequency of an allele is the number of occurrences of that allele in that population as a proportion of the number of occurrences of all the alleles of that gene. It is usually expressed as a decimal fraction, e.g. 0.6.

The sum of the frequencies of all the alleles of a gene must, therefore, equal 1.

Example

Imagine a population of 15 lizards, some green and some yellow. Skin colour is controlled by a single gene with two alleles: **G** (green) is dominant and **g** (yellow) is recessive.

- There are 15 individual lizards in the population, so there are 30 alleles of the gene for skin colour.
- Suppose six alleles in this population are **G** (green) and 24 are **g** (yellow).

The frequencies of these alleles are:
- 24/30 of the alleles are **g** — a frequency of **0.8**
- 6/30 of the alleles are **G** — a frequency of **0.2**

Examiner tip
You must write about the frequency of alleles and **not** the frequency of genes. The frequency of genes is always the same, since every diploid organism has two copies of each gene.

The Hardy–Weinberg principle

The Hardy–Weinberg principle states that the frequency of the alleles of each gene will not change from generation to generation.

This principle only applies, however, if the following conditions are met:
- The population is large.
- There is no movement of organisms into the population (immigration) or out of the population (emigration).
- The individuals reproduce sexually and mating is random.
- An individual's genotype does not affect its chances of breeding successfully.
- Gene mutation does not occur.

In practice, only the first of these conditions can be met in full. Others are not met — for example, in birds and mammals, there will be some kind of sexual selection, so mating is not random.

Examiner tip
Do not confuse the Hardy–Weinberg principle and the Hardy–Weinberg equation.

The Hardy–Weinberg equation

The Hardy–Weinberg equation enables us to estimate the frequency of different alleles and of genotypes in a population.

Assume there is a gene with two alleles: the dominant allele, **A**, and the recessive allele, **a**. In a population:
- the frequency of the dominant allele, **A**, is represented by p
- the frequency of the recessive allele, **a**, is represented by q
- there are no other alleles of this particular gene, so $p + q = 1$

We can use these allele frequencies to work out the frequencies of the three genotypes:
- The frequency of the dominant homozygote, **AA**, = p^2.
- The frequency of the recessive homozygote, **aa** = q^2.
- The frequency of the heterozygote, **Aa** = $2pq$.
- This accounts for all the possible genotypes, so: $p^2 + 2pq + q^2 = 1$. This is the Hardy–Weinberg equation.

If you know any of the values p, q, p^2 or q^2, you can calculate all the others.

Examiner tip
You will never be asked to explain how the Hardy–Weinberg equation is derived, so simply learn it so that you can apply it in an examination question.

Worked example 1

A dominant allele, **A**, has a frequency of 0.6 in a population. Calculate the frequency of the heterozygotes in the population.

Answer

We are told that $p = 0.6$

Since $p + q = 1$, $q = 1 - 0.6 = 0.4$

Therefore, the frequency of the heterozygotes, $2pq = 2 \times 0.6 \times 0.4 = 0.48$

Worked example 2

In a population, 64% of people can taste PTC. PTC tasting is determined by a dominant allele. Calculate the frequency of the dominant allele.

Answer

The PTC tasters include the dominant homozygotes (with a frequency p^2) and the heterozygotes (with a frequency $2pq$).

We cannot tell which of the tasters is homozygous and which is heterozygous. As is always the case, though, we can tell the homozygous recessives. We also know that their frequency (q^2) is, in this example, 0.36 (100% – 64% = 36%; as a decimal fraction this is 0.36).

So, $q^2 = 0.36$

$\therefore q = \sqrt{0.36} = 0.6$

and, because $p + q = 1$, $p = 1.0 - 0.6 = 0.4$

So the frequency of the dominant allele is 0.4.

If we know the size of a population, we can then work out actual *numbers* of each genotype:

> number in population × frequency of genotype

In example 2 above, if the population size was 450, then the number of individuals showing the recessive genotype is:

$0.36 \times 450 = 162$

Examiner tip

Whenever you answer a question involving the Hardy–Weinberg equation, look for the homozygous recessive individuals. You *know* their frequency is q^2 so you can easily calculate the values of q (as $\sqrt{q^2}$) and p (as $1 - q$).

Examiner tip

In calculations involving the Hardy–Weinberg equation, you must carry out the calculations using decimals. If you are given information about frequencies in percentages, change them to a decimal fraction (divide the percentage by 100) before you use it — for example, 64% = 64 ÷ 100 = 0.64.

Summary

After studying this topic, you should be able to:

- distinguish between the following terms: allele and gene; genotype and phenotype; heterozygous and homozygous
- use fully labelled genetic diagrams to predict the ratio of phenotypes resulting from monohybrid crosses involving: dominant and recessive alleles; codominant alleles; multiple alleles; alleles of sex-linked genes

- describe the concept of gene pool
- state the Hardy–Wenberg principle and the conditions under which the principle applies
- use the Hardy–Weinberg equation to find the frequencies of alleles or of genotypes from given data

Selection and speciation: the origin of new species

Selection

Key concepts you must understand

One of the key assumptions of the Hardy–Weinberg principle is that an individual's genotype does not affect its chances of breeding successfully. This is often not the case — one allele confers a distinct advantage over alternative alleles. As a result, the frequency of the favourable allele increases: this is **natural selection**. You learnt about this in the context of resistance to antibiotics in Unit 2. We can summarise natural selection as follows:

- A gene mutation produces an allele that confers an advantage.
- Organisms with this allele reproduce more successfully than those without it.
- As a result, they produce more offspring which, in turn, inherit the favourable allele.
- In this way, the frequency of the favourable allele increases.

Suppose that a feature is controlled by two alleles **A** and **a**. Initially, neither allele confers an advantage over the other and each allele has a frequency of 0.5. The environment changes and allele **a** gives a selective advantage to the organism, but allele **A** does not. The frequencies of the two alleles will change over time:

- The frequency of allele **a** will increase.
- The frequency of allele **A** will decrease (see Figure 57).

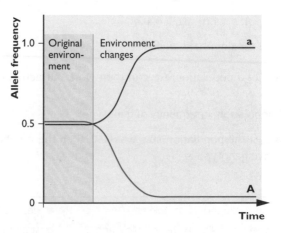

Figure 57 Change in allele frequency over time

Key facts you must know and understand

Some features show a range of values, rather than just the 'either/or' condition we have considered so far. The numbers of individuals at each point on the range reflect the survival value to the population of that particular value. The mean value represents the value that is best adapted to a particular environment.

AQA A2 Biology

Directional selection

If the environment changes, individuals at one extreme may have an advantage while those at the other extreme have a disadvantage. Over time, selection operates against the disadvantaged extreme and in favour of the other extreme. The mean and range of values shift towards the favoured extreme (see Figure 58).

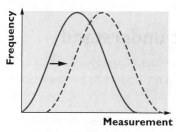

Environmental changes favour the selection of more suitable phenotypes, causing the normal distribution to shift

Figure 58 Directional selection

Example of directional selection

Peccaries feed on cacti. They eat those that have fewest spines. So, when peccaries enter an area where cacti are growing, they exert a 'selection pressure' on the cacti. Those with the most spines have an advantage; those with fewest spines are at a selective disadvantage. Those with the most spines survive and reproduce more effectively than those with fewer spines. Over time, the mean number of spines per cactus increases in the population.

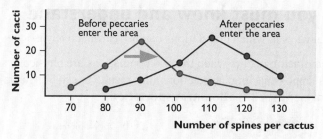

Figure 59 Directional selection in cacti

Stabilising selection

In a stable environment, selection operates against *both extremes* of a range. It operates to maintain the 'status quo' in the population and to make the population more uniform.

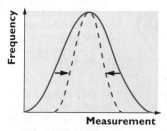

In a stable environment, environmental stresses tend to weed out unsuitable phenotypes, making the population more uniform

Figure 60 Stabilising selection

Birth mass in humans is an example. Babies who are very heavy or very light show a higher neonatal mortality rate (die more frequently at or just after birth) than those of medium mass. Over time, selection operates to reduce the numbers of heavy and light babies born.

Speciation

Key concepts you must understand

Natural selection explains how populations of a species become adapted to their environment and change in a changing environment. But how does a new species arise?

A key part of the definition of a species is 'a group of interbreeding organisms that produce fertile offspring'. If two groups cannot interbreed to produce fertile offspring, they must be different species.

As long as two populations are able to interbreed, they are unlikely to evolve into distinct species. They must somehow undergo a period of isolation — a period where they are prevented from interbreeding. During this period, mutations that occur in one population are not transferred to the other; so genetic differences between the populations can increase. These differences can become so great that, after some time, the populations are reproductively isolated, i.e. they cannot interbreed and are now distinct species.

Key facts you must know and understand

There are several ways in which populations can be reproductively isolated.

Geographical isolation occurs when the two populations are physically separated. Interbreeding is impossible and speciation may result. Speciation as a result of geographical isolation is called **allopatric speciation**.

Figure 61 Allopatric speciation

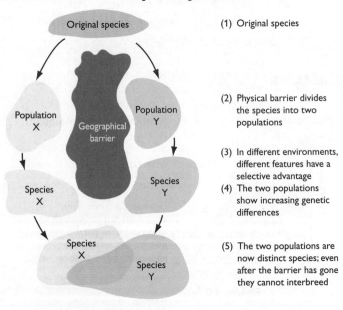

(1) Original species

(2) Physical barrier divides the species into two populations

(3) In different environments, different features have a selective advantage

(4) The two populations show increasing genetic differences

(5) The two populations are now distinct species; even after the barrier has gone they cannot interbreed

Other types of isolation need not involve physical separation. The two diverging populations may inhabit the same area, but be prevented from interbreeding in one of the following ways:

- Seasonal isolation — members of the two populations reproduce at different times of the year.
- Temporal isolation — members of the two populations reproduce at different times of the day.
- Behavioural isolation — members of the two populations have different courtship patterns.

Speciation following any of these methods of isolation is called **sympatric speciation**.

After studying this topic, you should be able to:

- use a familiar example and interpret unfamiliar information to explain how natural selection produces genetic change within a population
- distinguish between directional and stabilising selection
- explain how allopatric and sympatric speciation might occur

Summary

Questions & Answers

This section contains questions similar in style to those you can expect to see in the main body of **BIOL4**. The responses that are shown are real students' answers to the questions.

There are several ways of using this section. You could:

- 'hide' the answers to each question and try the question yourself. It needn't be a memory test — use your notes to see if you can actually make all the points you ought to make
- check your answers against the candidates' responses and make an estimate of the likely standard of your response to each question
- check your answers against the examiner's comments to see where you might have failed to gain marks
- check your answers against the terms used in the question — for example, did you *explain* when you were asked to, or did you merely *describe*?

Examiner's comments

Each question is followed by a brief analysis of what to watch out for when answering the question (shown by the icon ⓔ). All student responses are then followed by examiner's comments. These are preceded by the icon ⊖ and indicate where credit is due. In the weaker answers, they also point out areas for improvement, specific problems, and common errors such as lack of clarity, weak or non-existent development, irrelevance, misinterpretation of the question and mistaken meanings of terms.

Tips for answering questions

Use the mark allocation. Generally, 1 mark is allocated for one fact, concept or item in an explanation. Make sure your answer reflects the number of marks available.

Respond appropriately to the command words in each question, i.e. the verb the examiner uses. The terms most commonly used are explained below.

- **Describe** — you will find this in the final question in BIOL4, testing AO1, where it means 'write what you have learnt about'; in a question testing AO2 it means 'turn the trend shown by the data into words'.
- **Explain** — give biological reasons for *why* or *how* something is happening.
- **Calculate** — do some kind of sum and show how you got your answer.
- **Compare** — explain how the same feature is similar or different in two examples.
- **Complete** — add to a diagram, graph, flow chart or table.
- **Name** — rarely found in BIOL4; it means give the name of something identified in the question.
- **Suggest** — give a plausible biological explanation for something; this term is often used in questions testing understanding in an unfamiliar context, which is common in BIOL4.
- **Use** — you must find in the question, and include in your answer, relevant data or information.

Question 1 **Food chains and food webs**

The diagram shows a food web in decaying plant matter.

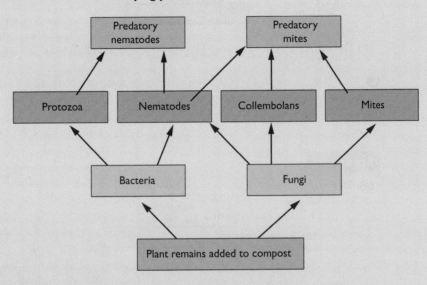

(a) For the food chain:

plant remains → bacteria → nematodes → predatory nematodes

construct:

(i) a pyramid of numbers (1 mark)

(ii) a pyramid of biomass (1 mark)

(b) A disease reduces the numbers of predatory nematodes. This could lead to a decrease in the breakdown of plant remains added to the compost. Explain how. (3 marks)

Total: 5 marks

ⓔ This is a straightforward question that you might find at the start of a BIOL4 paper. Part (a) tests your understanding of pyramids of number and of biomass. Part (b) requires you to work through a food web and see what effect a change in one trophic level has on other trophic levels. Notice that the last sentence contains the command word 'explain'.

Student A

(a) (i)

Predatory nematodes
Nematodes
Bacteria
Plant remains

(ii)

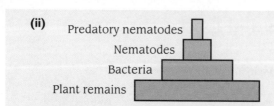

Predatory nematodes

Nematodes

Bacteria

Plant remains

(b) The bacteria and fungi won't break down the plant remains as fast **a**.

ⓔ **1/5 marks awarded** In (a)(i), this student does not seem to appreciate that bacteria are so small there must be more of them than there are plants. S/he draws a correct pyramid in (a)(ii), however, gaining 1 mark. **a** The answer to (b) simply re-states part of the question; it does not give reasons.

Student B

(a) (i)

Predatory mites   Predatory nematodes

Protozoa, nematodes   Mites

Bacteria and fungi

Plant remains

(ii)

Predatory nematodes

Nematodes

Bacteria

Plant remains

(b) Because there are fewer predatory nematodes, the numbers of protozoa and ordinary nematodes will increase **a**, because there is less predation. This means that they will consume more bacteria **a** and the nematodes will consume more fungi as well. So there will be fewer bacteria and fungi breaking down the plant remains **a**.

ⓔ **4/5 marks awarded** Student B has drawn an appropriate pyramid of numbers in (a)(i), but has labelled it incorrectly. S/he has drawn the pyramid correctly in (a)(ii) and gains 1 mark. In (b), this student clearly understands the relationships between the trophic levels and makes three valid points **a** in explaining the changes, scoring all 3 marks.

ⓔ **If asked to interpret unusual food webs like those in part (a)(i), make sure that you understand the principles of constructing ecological pyramids and, unlike student A, think about the size of organisms. Be careful not to make silly slips when you understand the biology, like student B did in (a)(i). Finally, when asked to explain something, make sure you give biological reasons; student A failed to gain any marks in (b) because s/he failed to offer any reasons at all. Student A gained 1 mark, a grade U performance, whereas student B gained 4 marks, a grade B performance.**

Question 2 Hardy–Weinberg

King cheetahs have a different pattern of spots from ordinary cheetahs. The king cheetah coat pattern is the result of a mutation. The resulting allele is recessive to that for normal coat pattern. A population of 100 cheetahs contained nine king cheetahs.

(a) At first, it was thought that the two might be different species. How could it have been proved that they were members of the same species? (2 marks)

(b) Use the Hardy–Weinberg equation to calculate:

   (i) the *frequency* of the dominant allele (2 marks)

   (ii) the *number* of heterozygotes in the population (2 marks)

(c) Give *two* reasons why the use of the Hardy–Weinberg equation might not be valid on this occasion. (2 marks)

Total: 8 marks

ⓔ Remember to start with the frequency of the homozygous recessive individuals when using the Hardy–Weinberg equation in part (b). They are the only genotype you can be sure of. In (c), you need to recall the conditions that must be met for the Hardy–Weinberg principle and, hence, the equation to be valid.

Student A

(a) They could breed them **a** and see if they produced offspring **b**.

(b) (i) There are 9 in 100 king cheetahs so this is 9% dominant alleles **c**.

   (ii) There are 91 ordinary cheetahs and probably two-thirds of these are heterozygous **d**. This is 60.66 cheetahs.

(c) The Hardy–Weinberg equation is only valid when there is a large population and 100 isn't that many **e**. The king cheetah is a mutation and they aren't supposed to happen in Hardy–Weinberg **e**.

ⓔ **3/8 marks awarded a** The student has the right idea, gaining 1 mark, but **b** forgets the importance of the offspring being fertile. **c** The student has calculated the percentage of the population that are king cheetahs, which was not asked for. **d** The student has shown no understanding of the Hardy–Weinberg equation and made a rash guess. **e** Although not well worded, these are two valid reasons and gain 2 marks.

Student B

(a) If a king cheetah and a normal cheetah are bred together and they produce offspring that can also have offspring **a**, then they must be the same species.

(b) (i) There are 9 in 100 = 9% king cheetahs. They have double recessive alleles so in Hardy–Weinberg this is q^2 **b**. So q^2 = 9% or 0.09 **c**.

So q = 0.3 and p = 0.7

(ii) Heterozygotes = $2pq$ **d** = $2 \times 0.3 \times 0.7$ = 0.42

So numbers of heterozygotes = 0.42×100 **e** = 42

(c) The Hardy–Weinberg principle is only valid for large populations and where there is no mutation.

ⓔ 8/8 marks awarded **a** This student realises the importance of breeding the two cheetahs and seeing if their offspring are fertile, scoring both marks. **b** The student has used information given in the question to work out that king cheetahs are homozygous recessive and that their frequency is q^2. **c** S/he has also remembered to convert a value of frequency into a decimal fraction. Having done so, getting the correct answer is straightforward. **d** The student has chosen the correct component from the Hardy–Weinberg equation and **e** calculated a number, rather than a frequency, as asked. In (c), s/he has identified two correct reasons.

ⓔ **Both students recalled the limitations of the Hardy–Weinberg principle. Although student A used clumsy wording, s/he still gained both marks for this answer. S/he showed little understanding of the use of the Hardy–Weinberg equation, however. Student A's score of 3/8 marks is borderline grade E; student B's score of 8/8 is a grade A performance.**

Question 3 **Aerobic respiration**

The diagram shows some of the stages in respiration.

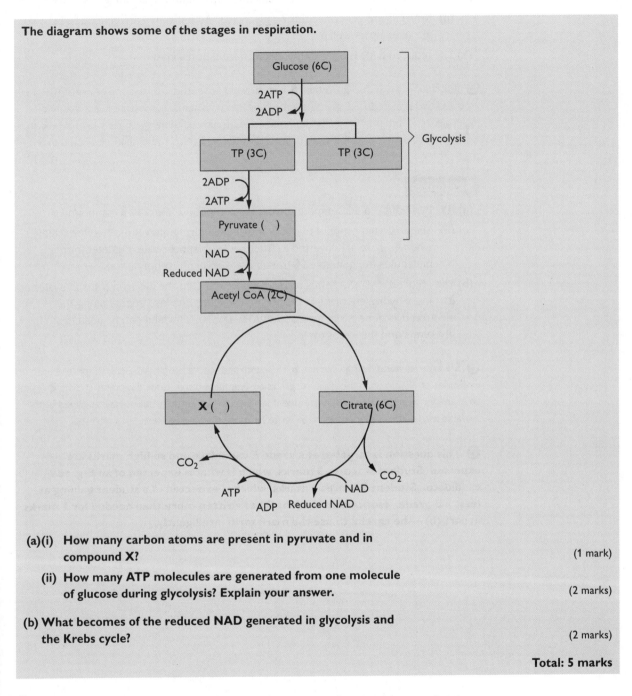

(a)(i) **How many carbon atoms are present in pyruvate and in compound X?**

(1 mark)

  (ii) **How many ATP molecules are generated from one molecule of glucose during glycolysis? Explain your answer.**

(2 marks)

(b) **What becomes of the reduced NAD generated in glycolysis and the Krebs cycle?**

(2 marks)

Total: 5 marks

This is a fairly typical question on respiration. At first sight, the diagram might seem daunting but it is there to help you to work out the answers to part (a). Part (b) tests pure recall (AO1), so you should find it easy.

Student A

(a) (i) Pyruvate has three carbon atoms and compound X has four **a**.

(ii) No ATPs are generated during glycolysis. Actually, two are generated but two are used up as well **b**.

(b) The reduced NAD is used in the electron transport **c** chain.

ℯ **3/5 marks awarded** **a** Correct. **b** The student has spotted that two molecules of ATP are used and two are produced, gaining 1 mark. S/he has, unfortunately, missed the fact that two molecules of TP are produced from each molecule of glucose. This answer gains 1 mark — **c** an examiner would not penalise the use of 'electron transport chain' rather than 'electron transfer chain'. The student has not mentioned that NAD is oxidised.

Student B

(a) (i) Pyruvate is a 3C compound whereas compound X is a 4C **a** compound.

(ii) There is a net profit of two ATP molecules during glycolysis. Two are used when glucose is converted to fructose 1,6-bisphosphate, but later four molecules are generated **b** by substrate-level phosphorylation.

(b) The reduced NAD (NADH) is re-oxidised **c** to NAD in the electron transfer system **d**. It loses hydrogen ions and electrons. The electrons lose energy along the electron transfer system to generate ATP. The NAD formed can be used again in glycolysis and the Krebs cycle.

ℯ **5/5 marks awarded** **a** Correct. **b** The candidate scores both marks, having spotted both molecules of TP shown in the diagram. **c** This answer gains 2 marks for the points shown. **d** Again, an examiner would not penalise the student for 'system' rather than 'chain'. Having gained both available marks, the student goes on to write far more than is required.

ℯ **This question is targeted at a grade E candidate and so high marks are expected. Student A scores 3 marks, which is what is expected of an E-grade candidate. Student B scores 5 marks, which is expected of a student gaining at least a B grade. Notice that student B has written more than needed for 2 marks in part (b) — be careful to use the mark tariff intelligently.**

Question 4 **Greenhouse effect**

The diagram shows the energy exchanges between the **Sun, the Earth and space in a situation where the average temperature of the Earth is stable at 14°C. (Figures for energy exchange are watts per square metre of the Earth's surface.)**

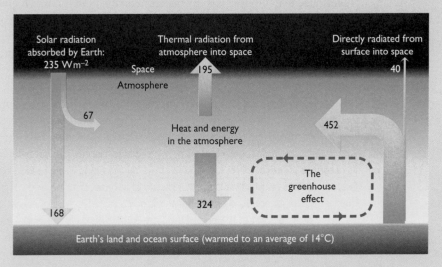

Solar radiation absorbed by Earth: 235 Wm^{-2}

Thermal radiation from atmosphere into space: 195

Directly radiated from surface into space: 40

Space
Atmosphere

67

Heat and energy in the atmosphere

452

The greenhouse effect

168

324

Earth's land and ocean surface (warmed to an average of 14°C)

(a) Name two greenhouse gases.   (2 marks)

(b) Use figures from the diagram to explain why the surface of the Earth stays at a constant 14°C. (4 marks)

(c) Use figures from the diagram to explain what would happen to the temperature of the Earth if there were no greenhouse effect.   (4 marks)

Total: 10 marks

ⓔ Part (a) tests AO1 and does not require you to use the diagram. Notice that parts (b) and (c) of this question require you to use evidence from the diagram.

Student A

(a) Carbon dioxide and ozone **a** are greenhouse gases.

(b) It stays the same temperature because it is losing as much heat as it gains **b**.

(c) If there was no greenhouse effect then global warming wouldn't take place. The planet would be cooler **b** and the ice caps wouldn't be melting.

ⓔ **3/10 marks awarded** Carbon dioxide is correct and scores 1 mark but **a** ozone is not a greenhouse gas. **b** In both answers, the student gains 1 mark for stating the general principle but has made no attempt to use figures from the diagram, as instructed.

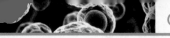

Student B

(a) Carbon dioxide and methane **a**.

(b) Total energy absorbed by the Earth's surface is 168 + 324 = 492 **c**. To stay the same temperature, it must lose the same amount of heat **b**.

(c) The Earth's temperature would be lower if there was no global warming. It's the greenhouse effect that keeps the planet habitable. But too much greenhouse effect is causing global warming and melting the ice caps. This means that the oceans will rise and some lands will be flooded. If all the energy reflected back to Earth by the greenhouse effect was lost to space, the Earth would cool down rapidly. $452\,Wm^{-2}$ is lost **d** in the diagram.

ⓔ 5/10 marks awarded **a** This answer scores both marks. **b** The student gains 1 mark for stating the general principle and **c** 1 mark for using figures from the diagram. S/he has not explained how the values balance — the $492\,Wm^{-2}$ that are gained are balanced by the $452\,Wm^{-2}$ lost to the atmosphere and $40\,Wm^{-2}$ radiated directly into space. In (c), this student commits a serious examination error — s/he strays off the point into a related area, but one that will score no marks. S/he must have spent a couple of minutes thinking, and writing, about the consequences of the greenhouse effect, but this is not required by this question. Make sure that you *stick to the point*. **d** Right at the end of the answer, s/he makes a superficial use of the figures and scores 1 mark.

ⓔ **Student A scores 3 marks — a borderline E performance; student B scores 5 marks — a grade D performance. Reading the questions carefully is always important, but particularly so here. The instruction in parts (b) and (c) to 'use the figures' means that there is a huge penalty if you do not.**

Question 5 **Nitrogen cycle**

The diagram represents the circulation of nitrogen-containing compounds in an aquarium. High concentrations of nitrates are toxic to many animals.

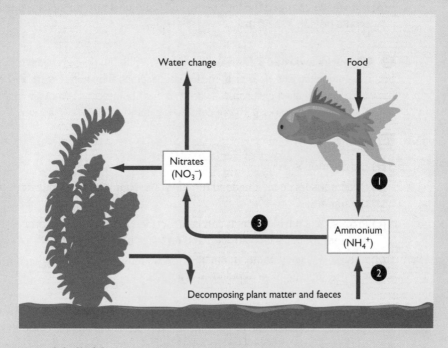

(a) Name the process labelled 1. (1 mark)

(b) Describe how ammonium ions are formed in the process labelled 2. (3 marks)

(c) Name the process labelled 3 and explain why it can be described as oxidation. (2 marks)

(d) Explain why the water in the aquarium should be changed regularly. (2 marks)

Total: 8 marks

The first part of a question is often straightforward and this is no exception; you should not be troubled by (a). Think carefully about the process involved in (b) and take note of the mark tariff before answering, You need to have an understanding of the chemistry of oxidation–reduction to answer (c); if you do not, don't worry — just move on to part (d). In answering part (d), use information from the diagram rather than inventing something.

Student A

(a) Excretion **a**

(b) The dead remains are decomposed **b**.

(c) Process 3 is denitrification. It's an oxidation process because it uses oxygen **c**.

(d) If it wasn't changed the nitrates would build up **d** and the question tells us this would poison **d** the fish.

e **4/8 marks awarded a** This answer is acceptable — fish do excrete ammonia, which would become ammonium ions in water. **b** The idea of decomposition gains 1 mark. **c** The student has incorrectly identified the process as denitrification but gives too simplistic an answer about oxidation to gain any marks. **d** The student gains 2 marks for these two answers.

Student B

(a) Defaecation **a**

(b) Decómposers decay the dead remains. The ammonium comes from proteins **b** in the remains.

(c) Process 3 is nitrification. It can be described as an oxidation process because the ammonium loses its hydrogen and replaces it with oxygen **c**.

(d) If the water isn't changed, nitrification will use up all the oxygen in the water and the fish will die from asphyxiation **d**.

e **5/8 marks awarded a** This answer is acceptable. **b** The student scores 2 marks for the concept of decomposition and the statement that proteins are a source of the nitrogen that forms ammonium ions. In (c), the student correctly identifies nitrification and **c** uses information in the diagram to show understanding of the oxidation process. Unfortunately, in (d), the candidate has ignored the information given in the stem of the question. **d** Although s/he comes up with an idea s/he thought reasonable, any oxygen lost in nitrification would be replaced by diffusion from the atmosphere.

e **Student A scores 4 marks (grade C) and student B scores 5 marks (also grade C). Many candidates confuse the names of processes in the nitrogen cycle, yet there aren't that many to remember. There is no short cut — you just have to learn them. Only student A made use of information in the stem about nitrates poisoning fish. Don't forget, in questions testing your ability to apply understanding to unfamiliar contexts (which is what most of the questions in BIOL4 do), examiners give you information in the stem that you cannot be expected to know but need to use when answering parts of the question.**

Question 6 **Inheritance**

Andalusian fowl can have plumage with three distinct colours:
- **black**
- **white**
- **blue**

In breeding experiments, the following results were obtained:

Parents	black × white
Offspring	all blue

Parents	blue × blue
Offspring	black : blue : white
	1 : 2 : 1

(a) **Suggest an explanation for these results. Use evidence from the crosses to support your explanation.**   (5 marks)

(b) **If a blue fowl were bred with a white fowl, what offspring would you expect? Explain your answer.**   (3 marks)

Total: 8 marks

ⓔ Most students like straightforward genetics questions like this and do them well. Genetics questions will almost always use a context that is unfamiliar to you but you should use the format of a labelled genetic diagram, with which you are familiar, to structure your answer.

Student A

(a) It looks like codominance **a**. Blue is a new colour produced by parents who are black and white. If the black parent had two genes for black and the white parent had two genes for white, the children could have one gene for black and one gene for white **b**, which would produce blue.

(b) If you bred blue and white together, you would probably get the **c** same again. You couldn't get black offspring because that would need two genes for black **d**.

ⓔ **2/8 marks awarded a** The student clearly recognised that codominance was involved but there is no mark for simply stating this — the marks are for an explanation. **b** The student has not used a genetic diagram, but has effectively given the genotype of the black and white parents in cross 1 and both the genotype and phenotype of all their offspring, gaining 2 marks. Notice the student has incorrectly written about genes when s/he meant alleles (of the gene for feather colour) and naively written about the offspring as children. **c** It is not clear what 'same again' means — the same as what? **d** The student helps a little by writing that you would not get black offspring but fails to tell us in what ratio we should expect blue and white offspring. Again, an explanation is needed in order to gain marks.

Student B

(a) If the parents are homozygous for black and white, they could have blue offspring if black and white were codominant.

Genotype of parents	C^BC^B × C^WC^W **a**	
Gametes	C^B C^W	
Genotype of offspring	C^BC^W **b**	
Phenotype of offspring	all blue (which is what the question shows)	

Two blue parents could produce a mixture of offspring:

Genotype of parents	C^BC^W × C^BC^W	
Gametes	C^B C^W C^B C^W **c**	
Genotype of offspring	C^BC^B C^BC^W C^BC^W C^WC^W **d**	
Phenotype of offspring	black blue blue white (a 1:2:1 ratio as in the question) **e**	

(b)

Genotype of parents	C^BC^W × C^WC^W	
Gametes	C^B C^W C^W **f**	
Genotype of offspring	C^BC^W C^WC^W **g**	
Phenotype of offspring	blue white	

There would be equal numbers of blue and white offspring. **h**

ⓔ **8/8 marks awarded** The student has done what a teacher will have coached her/him to do and used labelled genetic crosses. Points **a–h** show where s/he gained all 8 marks for this question.

ⓔ **Despite showing some understanding of inheritance, student A scores just 2 marks — a low grade U. Student B has done what all candidates should do — used labelled genetic diagrams that provide a full explanation. Although student B has shown grade A performance on this question, examiners regularly see students across all grade ranges for the final BIOL4 result gaining full marks on questions like these.**

Question 7 Energy and food production

Organisms that reduce the yield of a crop plant are called pests. They can be controlled using pesticides, by biological control or by integrated crop management.

The diagram shows the effects of repeated pesticide applications on a population of pests.

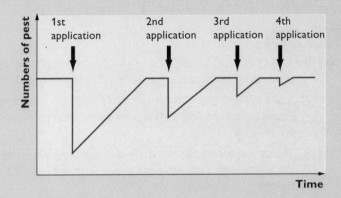

(a) (i) Suggest two reasons why the first application of the pesticide does not reduce the pest population to zero. (2 marks)

 (ii) Explain the reduction in effect of the pesticide at the second, third and fourth applications. (3 marks)

(b) Give one benefit of each of the following in an integrated crop management system:

 (i) crop rotation (not growing the same crop in the same field in successive years)

 (ii) using organic fertilisers (such as farmyard manure) rather than inorganic fertilisers

 (iii) planting crops that are tolerant of the local soil pH conditions (3 marks)

Total: 8 marks

ⓔ Part (b) of this question is a test of recall (AO1) and does not involve using the diagram. If you feel more confident about your AO1 skills than your AO2 skills, you might choose to answer part (b) first. Note that the mark tariff means 1 mark for each benefit. Part (a) involves interpretation of a graph (AO2). Note that (a)(i) uses the command word 'suggest'; you should know why examiners have used that word.

Student A

(a) (i) The dose might not have been strong enough **a** and some of them might not have **a** been affected by it.

 (ii) More of the pests become **b** immune to the pesticide. These survive and those that aren't immune are killed and so there are less of them.

(b) (i) Insect pests won't have the same crops to feed on **c** and so they die out.

 (ii) To supply what the plant needs in a natural form **d**.

 (iii) If they weren't from the area, and the soil was too acid, they might not grow properly **e**.

ⓔ **3/8 marks awarded a** What does this student mean by 'not strong enough' and 'not affected by it'? Both are too vague to gain a mark. **b** S/he then writes about immunity, which is wrong — you become immune to antigens, not pesticides. There is a hint, however, that as time goes by, there will be more of the resistant forms in the population and less of the non-resistant forms and this scores 1 mark. **c** This answer gains 1 mark. **d** This answer is a good example of the sort of thing you should *never* write. It conveys about as much biological information as saying vegetables have a lot of 'goodness' in them. Be precise and give answers that show you have learnt something from your A2 biology course. **e** This answer gains 1 mark.

Student B

(a) (i) Some of the pests might have a mutation that gives them resistance **a** to the pesticide. Insecticides are usually sprayed and not all the pests are 'hit' **a**.

(ii) Some of the pests are resistant and these survive the initial application. **b** These reproduce so that when the insecticide is re-applied, fewer of the pests are killed **b**.

(b) (i) The pests probably feed mainly on one crop plant, so by changing the crop every year you don't get a build up of any one pest. **c** It might not kill them off though.

(ii) Organic fertilisers don't supply all the minerals all at once; **d** they release them slowly, like a 'drip-feed'.

(iii) They will be able to compete effectively with any local weeds, **e** which will also be adapted to the conditions.

ⓔ **7/8 marks awarded a, b** Each of these points gains 1 mark. **c, d, e** Each of these benefits is clear, valid and gains the available 1 mark.

ⓔ **Student A has given answers that lack the precise use of terminology you should demonstrate at A-level. S/he gains 3 marks — a grade D performance. In contrast, student B has learnt and used biological terminology and gains 7 marks — a grade A performance. The hardest part of the question was (a)(ii) in which neither student referred to fewer pests being killed at each repeated application.**

Question 8 **Selection and speciation**

The graph shows the distribution of root length in a population of grass. The population inhabits an area in which the soil water is held mainly in the top 20 cm.

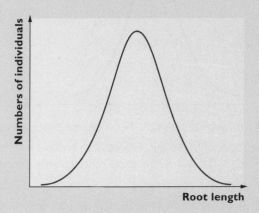

(a) **What does the term 'population' mean?** (1 mark)

(b) (i) **Sketch, on the graph, the distribution of root lengths you would expect if some of these plants now colonised a different area where the soil water was held mainly below 20 cm.** (1 mark)

(ii) **Name, with a reason, the type of selection operating in this example.** (1 mark)

(iii) **Describe the evolutionary mechanisms that would lead to this change in the distribution of root lengths.** (4 marks)

(c) **In time, these populations might evolve into different species.**

(i) **Would this be an example of sympatric or allopatric speciation? Explain your answer.** (1 mark)

(ii) **Describe and explain the conditions essential for speciation to occur.** (2 marks)

Total: 10 marks

ⓔ Watch out for questions like part (a) — here examiners want a precise definition of this ecological term. Notice that parts (b)(ii), (c)(i) and (c)(ii) require you to give reasons; simply hitting on the correct name in (b)(ii) and (c)(i) or giving a description in (c)(ii) would not get you the available marks.

Student A

(a) A group of individuals **a** living in the same place at the same time.

(b) (i)

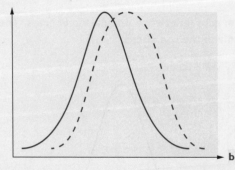

(ii) Directional selection, because one extreme has an advantage over the other **c**.

(iii) Those with the longer roots can survive better **d** and breed because they can reach the water. The short-rooted plants die out **e** by natural selection.

(c) (i) Allopatric speciation **f**

(ii) To become new species, the two populations must become increasingly different. They won't be able to breed **g**.

ℯ **5/10 marks awarded a** The student has failed to mention that the individuals are of the same species. **b** This graph shows the whole distribution moving and gains 1 mark. **c** Correct. The student gains 1 mark each for **d** and **e**. **f** The student has chosen the correct term but has not given the required reason and fails to score. **g** Mentioning inability to interbreed gains 1 mark.

Student B

(a) A population is a group of the same species **a** living in the same place.

(b) (i)

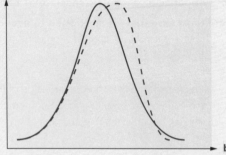

(ii) Directional selection, because selection favours the longer-rooted plants as they can obtain the deep water more easily **c**.

(iii) The longer-rooted plants have a selective advantage **d** because their roots can obtain water more efficiently than those with short roots. So the long-rooted plants survive and reproduce **d** and pass on their 'long root' genes to their offspring **d**. This keeps happening each generation so there are more and more long-rooted plants as time goes by.

(c) (i) Allopatric speciation because they do not interbreed **e**.

    (ii) The two populations must not interbreed **f**, so that they become different species.

🅔 6/10 marks awarded **a** The student has identified a population comprising a single species. **b** Unfortunately, the student has not shown the whole of the distribution shifting and shows the same plants with short roots. **c** The term is correct and the reason is sound. **d** The candidate scores 3 marks for these three valid points. **e** Although the student gives the correct term, s/he does not indicate that the populations cannot interbreed because they are geographically isolated. **f** This answer gains 1 mark.

🅔 **Student A scores 5 marks and student B scores 6 marks; both represent a grade C performance. Both students failed to gain marks by providing insufficient detail in their answer — you should use the mark tariff to indicate how many points to make. Neither student involved gene mutations to explain the origin of plants with even longer roots in (b)(iii). Similarly, in (c)(ii), neither gave a full explanation of speciation, which should include: populations being reproductively isolated; gene pools becoming increasingly different as a result of this isolation; the difference in gene pools leading ultimately to an inability to interbreed.**

Question 9 Populations and succession

The diagrams show a primary succession originating in a pond. The graph shows the accompanying changes in biomass, primary production and species diversity.

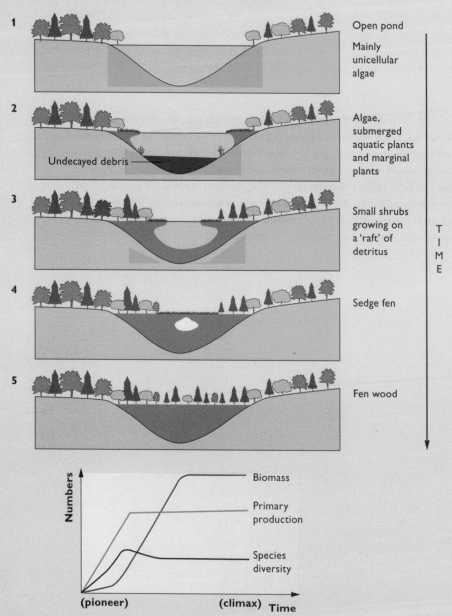

(a) Use the graph and your knowledge of succession to explain the changes occurring in the first three stages. (5 marks)

(b) How could you collect the data needed to calculate the species diversity at stage 4? (5 marks)

Total: 10 marks

ⓔ Examiners cannot expect you to know any particular example of succession; this is why you have been given so much information about this particular example. You should be able to apply your understanding of succession to this, or any other, example given sufficient detail in the question. Part (b) does not ask you to use data to calculate the index of diversity. Instead, you are given a question testing your AO3 skills relating to section 3.4.1 of the specification. From the wording of this part of the specification, examiners can expect you to have carried out an ecological investigation and used these techniques. Take note of the high mark tariff for the two parts of this question as you prepare your answers.

Student A

(a) As time goes by, more and more different plants start to appear **a** in the pond. Eventually these die and decay and add to the nutrients **b** in the pond. This means that even more plants can survive. Eventually the pond gets filled in with the dead remains of the plants and a fen wood grows on top of it.

(b) You would need to throw some quadrats at random **c** and count the number of organisms of each type **d** in the quadrat. If you got an average, you can calculate the number of each in the area **e** and use this to calculate the species diversity.

ⓔ **4/10 marks awarded** In part (a), the student has ignored the instruction to refer only to the first three stages. S/he made only two valid points. This can be interpreted as **a** more species of plant can establish in the changed environment, with **b** a specific change in the environment given. **c** Throwing quadrats at random will never gain credit. **d** Having failed to gain a mark for throwing quadrats, the student gains a mark for stating correctly that s/he would count the organisms of each type (species) within each quadrat. **e** Finally, s/he gives an acceptable description of how to estimate the abundance of different species in a given area (though a unit of volume would be appropriate in a pond).

Student B

(a) The original algae die and decay and add nitrates **a** to the pond water. This allows more and different species of plants to colonise **b** the pond. Some of the dead remains of plants float on the water and provide a new habitat for other plants and small shrubs that start to appear.

(b) To calculate the species diversity, you need to calculate a species diversity index. The formula for this index is $d = (N(N - 1))/(\Sigma n(n - 1))$.

You need to know how many there are of each species **c** in the area. You can actually count the number of big trees **d**, but for small plants and animals, you need to estimate the size of the population. You can use the mark–release–recapture technique to estimate the size of animal populations **e** and random quadrats to estimate the size of plant populations. Use a random number generator and place the quadrats at the coordinates it generates **f**. Count the number of individual plants in each quadrat and scale up to the whole area **g**. You can then use the formula to calculate the diversity index.

ⓔ **7/10 marks awarded** In **a** and **b**, this student makes the same two points as student A and gains the same 2 marks. S/he makes the same two points again in the third sentence; in a question like this, try to ensure you make five separate points for 5 marks. **c** This student makes a better attempt at part (b) and gains 1 mark for stating clearly what s/he needs to find. S/he gains a

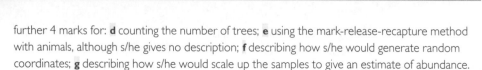

further 4 marks for: **d** counting the number of trees; **e** using the mark-release-recapture method with animals, although s/he gives no description; **f** describing how s/he would generate random coordinates; **g** describing how s/he would scale up the samples to give an estimate of abundance.

🅮 **Disappointingly, neither student has used the graph at all, despite the clear instruction to do so in the wording of part (a). Both students have given a superficial account of succession, for example neither refers to the concept of pioneer species (the unicellular algae) or to interspecific competition, and neither makes a clear reference to the fact that decomposition is partial, resulting in a build-up of plant material that eventually fills the pond. Both have, however, expressed the idea of the environment being altered (by increasing ion concentration). Don't forget that you can be asked about experimental techniques in BIOL4, like the question in part (b). Overall, student A scores 4 marks (grade E) and student B scores 7 marks (grade B).**

Question 10 Photosynthesis and energy transfer

(a) The flow chart summarises the fate of light energy striking a leaf.

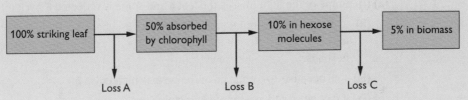

- **(i)** Give two processes that could contribute to loss **A**. (2 marks)
- **(ii)** What does loss **B** tell you about the efficiency of the process of photosynthesis? Explain your answer. (2 marks)
- **(iii)** In this example, what percentage of light energy becomes part of the net primary production? Explain your answer. (3 marks)

(b) The diagram below summarises the main events during the light-independent reaction of photosynthesis. The graph shows changes in the levels of RuBP and GP in a chloroplast when the light source is removed.

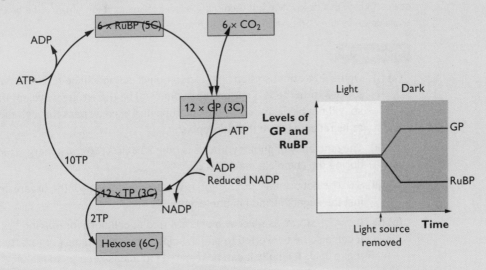

- **(i)** Give two possible fates of the hexose produced. (2 marks)
- **(ii)** Use the diagram to explain the changes shown in the graph in the levels of RuBP and GP in the chloroplast when the light source is removed. (6 marks)

Total: 15 marks

ⓔ In (a)(i), the command word 'give' means you should just write the appropriate two names and move on; do not write a long answer. Don't, however, rush into (a)(i); read the stem and the first box in the diagram carefully before answering. Note that parts (a)(ii) and (a)(iii) do not require you to perform a calculation. You should find part (b)(i) very easy but (b)(ii) will require a bit more thought. You need to think about what is needed to produce RuBP to answer this successfully.

Try to work out what is going on — making a list in the margin if it helps — before writing your answer.

(a) (i) Some of the light misses the leaf **a** and some is reflected **b** back into space.

  (ii) It shows that it is quite an inefficient process, because there is a considerable loss **c** of energy. Not all the energy absorbed by chlorophyll ends up in hexoses.

  (iii) 10% because 10% ends up in the hexose molecules **d**.

(b) (i) It can be stored as glucose **e** or starch.

  (ii) When the light source is removed, RuBP decreases and GP increases. This is because **f** RuBP is converted to GP in the light-independent reaction. When all the RuBP has been used up, no more GP can be made.

ⓔ **4/15 marks awarded a** This student has not read the question; s/he is told (twice) that the light has struck the leaf. **b** I mark is gained for stating that some light is reflected. **c** S/he gains I mark for realising that energy is lost but fails to explain why. **d** S/he has not understood the concept of net primary productivity and fails to gain a mark in (a)(iii). **e** In (b)(i), s/he gains I mark for naming starch but seems to have forgotten from Unit I that glucose is a hexose. The student starts (b)(ii) with a description but then provides an explanation **f** for I mark. S/he provides no explanation for the observation in the final sentence of the answer.

(a) (i) Of the 50% not absorbed by the chlorophyll, some will be reflected back into space **a** from the leaf surface and some will be the wrong wavelength **a** and so will not be absorbed by the chlorophyll. This may pass through the leaf **a** or be reflected from the chloroplast.

  (ii) The efficiency of photosynthesis is only 20% (10%/50%) as energy is lost **b** during the chemical reactions **c**.

  (iii) 5%. The net primary production is the actual biomass of the organism, **d** not just the sugars formed in photosynthesis.

(b) (i) It can be stored as glucose **e** or converted to cellulose for making cell walls **f**. It can also be converted to amino acids to make proteins **f** or to fatty acids to make lipids **f**. Finally it can be respired **f** to release energy to make ATP.

  (ii) In the light-independent reaction, RuBP is converted into GP, which is then converted to TP. Some of this is re-converted back **g** into RuBP. This needs ATP and reduced NADP **h** from the light reactions **i**. When the light is removed, the ATP and reduced NADP cannot be made **j**. Therefore, the RuBP cannot be formed again, so the level falls **k**.

ⓔ **13/15 marks awarded** Rather than give two processes, as asked in (a)(i), this student has given **a** three answers. This is a risky strategy; if any were wrong it would cancel out a mark given for a correct answer. In this case, none is incorrect and the student gains 2 marks. S/he gains I

mark for **b** and another for explaining where the loss occurs **c**. S/he gives the correct value in (a)
(iii) and provides a partial explanation of net primary productivity **d**; s/he forgets to mention
respiratory losses. As with her/his answer to (a)(i), this student provides more answers **f** than
are asked for. This time, one is incorrect **e** and cancels the mark given for one of the correct
answers. Since s/he has given four correct answers, however, even after cancelling s/he still gains
the maximum 2 marks. In (b)(ii), s/he explains that, for TP to be converted to RuBP **g**, ATP and
reduced NADP **h** from the light-dependent reaction **i** are used. In the absence of light, these
cannot be made **j** and so the level of **k** RuBP falls. Note that the examiner has not penalised her/
him for the reference to the light reactions **i**; it is clear what 'light reactions' means.

**ⓔ Student A scores only 4 marks, which is a borderline E/U response. Student B
has read the question more carefully and used information in the question and her/
his own knowledge to provide a good explanation in (b)(ii). S/he gains 13 marks,
which is a grade A response.**

Knowledge check answers

1 It is a place where a community of organisms (bacteria and fungi) is found.

2 Interspecific competition; light

3 (a) The predator population responds to changes in the availability of food. (b) The predator does not eat all the available prey/energy is lost between trophic levels of food chains.

4 (a) A biotic factor, such as the amount of food/oxygen. (b) Interspecific competition.

5 When a running mean of abundance or number of species in each quadrat becomes constant.

6 A belt transect — you are investigating the effect of a variable that will change as you move away from the motorway.

7 (a) 40 (b) Population size did not change (no births, deaths or migration); the marked snails dispersed into the population.

8 Use of soap, since this is followed by the steepest fall in death rate.

9 (a) It is low. (b) It is lower than male mortality at all ages.

10 Molecules are reduced by the addition of electrons and oxidised by the loss of electrons.

11 (a) ATP and reduced NADP (b) Oxygen

12 Energy from ATP; hydrogen (electrons) from reduced NADP

13 It has a higher temperature (enzyme action) and a higher concentration of CO_2.

14 It does not increase the rate of photosynthesis (growth) but costs money; eventually a high temperature denatures enzymes.

15 (a) It has a greater surface area for the electron transfer chain. (b) Reactions of the Krebs cycle occur more rapidly in solution.

16 Too large a molecule to pass through the membrane/no glucose transporter proteins in the outer mitochondrial membrane

17 (a) 2 molecules of ATP (b) Lost from triose phosphate

18 A phosphate group is transferred from an organic molecule (e.g. TP) to ADP.

19 ADP and P_i combine to form ATP; oxygen is the final electron acceptor.

20 2 ATP molecules are used to phosphorylate glucose, but 4 molecules are regenerated

21 Some might be the wrong wavelength, be reflected, pass through the leaf or not strike a chloroplast.

22 (a) 83 240/6 800 000 = 1.2%; (b) 1532/13 472 = 11.4%

23 Cellulose/lignin (in plant cell walls)

24 $NP = I - (R + U + F)$

25 Carnivores do not eat many of the herbivores; dead herbivores and faeces are decomposed.

26 Improves soil structure; improves water retention of soil; less risk of eutrophication

27 Look back to Figure 3 — as prey numbers decline, so do predator/parasite numbers.

28 Diet – increases I and reduces F; restricting movement and keeping indoors both reduce R.

29 Dead plants/plant leaves; dead animals; animal faeces

30 Seasonal changes in temperature and light intensity affect the rate of photosynthesis.

31 Prior to 1925, rise in CO_2 concentration continuous but temperature shows fluctuations; temperature could be fluctuating about a mean value.

32 The two curves show a correlation; changes in temperature sometimes occurred before changes in CO_2 concentration and sometimes after changes in CO_2 concentration.

33 Nucleotides/nucleic acids

34 (a) More evaporation means concentrations of ions increase/higher temperature increases enzyme activity. (b) Movement re-oxygenates the water/dilutes the solution of ions.

35 Light (they are taller and shade grasses); water or mineral ions (shrubs have deeper or more widespread root systems than grass).

36 About 20 years — before productivity decreases and gaps appear in the heather.

37 (a) Tt (b) T and t

38 Expectation: equal numbers of plants with pink flowers and plants with white flowers.
Explanation:

Phenotype of parents	Pink flowers	White flowers
Genotype of parents	C^RC^W	C^WC^W
Gametes		
Genotype of offspring	C^RC^W	C^WC^W
Phenotypes of offspring	Pink flowers	White flowers
Ratio of phenotypes	1:1	

39 No. If the mother's genotype is I^AI^O and the father's genotype is I^BI^O, there is a 1 in 4 chance of a child of genotype I^OI^O (group O).

40 She must be X^HX^h, a carrier, so expect half of her eggs to carry the X^h allele.

41 (a) Continuous variation (b) It is controlled by more than one gene/is polygenic.